Electrochemical Capacitors

Electrochemical Capacitors

Editors

Seiji Kumagai
Daisuke Tashima

MDPI • Basel • Beijing • Wuhan • Barcelona • Belgrade • Manchester • Tokyo • Cluj • Tianjin

Editors

Seiji Kumagai
Department of Mathematical Science and
Electrical-Electronic-Computer Engineering
Japan

Daisuke Tashima
Department of Electrical Engineering, Fukuoka Institute of Technology
Japan

Editorial Office
MDPI
St. Alban-Anlage 66
4052 Basel, Switzerland

This is a reprint of articles from the Special Issue published online in the open access journal *Batteries* (ISSN 2313-0105) (available at: https://www.mdpi.com/journal/batteries/special_issues/ Electrochemical_Capacitors).

For citation purposes, cite each article independently as indicated on the article page online and as indicated below:

LastName, A.A.; LastName, B.B.; LastName, C.C. Article Title. *Journal Name* **Year**, *Article Number*, Page Range.

ISBN 978-3-03936-722-1 (Pbk)
ISBN 978-3-03936-723-8 (PDF)

Contents

About the Editors

Seiji Kumagai (Ph.D.) is a professor of electrical and electronic engineering at Akita University, Akita, Japan. He received his Ph.D. from Akita University in 2000 and experienced a postdoctoral research fellow at the Japan Society for the Promotion of Science from 2000 to 2002. He worked at Akita Prefectural University from 2002 to 2011 as an assistant professor and associate professor. Since 2011, he has served at Akita University. His research interests include electrochemical energy storage materials and devices such as capacitors and batteries.

Daisuke Tashima (Ph.D.) is an associate professor of electrical engineering at Fukuoka Institute of Technology, Fukuoka, Japan. He received his B.E degree in Electrical and Electronic Engineering from the University of Miyazaki in 2003, and his doctoral degree from the University of Miyazaki in 2007. He is now an associate professor at the Department of Electrical Engineering, Fukuoka Institute of Technology. His research interests include electric double-layer capacitors and proton exchange membrane fuel cells.

Preface to "Electrochemical Capacitors"

Technological development of energy storage devices is now ongoing worldwide for mitigating carbon dioxide emissions to combat global warming. Several industrially promising sectors such as automobiles, renewable energies, power quality management, and mobile gadgets require high-performance and low-cost energy storage devices. Electrochemical capacitors have been accepted as the key elements for realizing charge–discharge cycling with high power density, high efficiency, and long life. Electric double-layer capacitors, pseudocapacitors, and hybrid capacitors, which can be called supercapacitors, are typical modern electrochemical capacitors intended for energy storage. To record the cutting-edge knowledge and the latest experience related to electrochemical capacitors, five high-quality papers contributing to further technological developments in electrochemical capacitors have been selected for this book. The effect of the milling degree of activated carbon particles used in the electrodes on the supercapacitive performance of an electric double-layer capacitor is discussed, providing practically useful knowledge for manufacturing the electrodes of electric double-layer capacitors. Supercapacitive performances of nickel molybdate/reduced graphene oxide nanocomposite, copper-decorated carbon nanotubes, and nickel hydroxide/activated carbon composites are described to find more promising electrode materials for electrochemical capacitors. Hybrid utilization of electrochemical capacitors with other types of energy devices (photovoltaics, fuel cells, and batteries) in DC microgrid is also reported. This book will be beneficial for researchers and engineers in the field of electrochemical capacitors.

Seiji Kumagai, Daisuke Tashima
Editors

batteries

Article

Effect of Ball Milling on the Electrochemical Performance of Activated Carbon with a Very High Specific Surface Area

Takuya Eguchi [1], Yugo Kanamoto [1], Masahiro Tomioka [1], Daisuke Tashima [2] and Seiji Kumagai [1,*

[1]   Department of Mathematical Science and Electrical-Electronic-Computer Engineering, Akita University, Tegatagakuen-machi 1-1, Akita 010-8502, Japan; d8519003@s.akita-u.ac.jp (T.E.); m8020409@s.akita-u.ac.jp (Y.K.); tomioka@gipc.akita-u.ac.jp (M.T.)

[2]   Department of Electrical Engineering, Fukuoka Institute of Technology, Wajiro-higashi 3-30-1, Higashi-ku, Fukuoka 811-0295, Japan; tashima@fit.ac.jp

*   Correspondence: kumagai@gipc.akita-u.ac.jp; Tel.: +81-18-889-2328

Received: 10 February 2020; Accepted: 10 April 2020; Published: 14 April 2020

Abstract: Activated carbon (AC) with a very high specific surface area of >3000 $m^2\ g^{-1}$ and a number of course particles (average size: 75 µm) was pulverized by means of planetary ball milling under different conditions to find its greatest performances as the active material of an electric double-layer capacitor (EDLC) using a nonaqueous electrolyte. The variations in textural properties and particle morphology of the AC during the ball milling were investigated. The electrochemical performance (specific capacitance, rate and cyclic stabilities, and Ragone plot, both from gravimetric and volumetric viewpoints) was also evaluated for the ACs milled with different particle size distributions. A trade-off relation between the pulverization and the porosity maintenance of the AC was observed within the limited milling time. However, prolonged milling led to a degeneration of pores within the AC and a saturation of pulverization degree. The appropriate milling time provided the AC a high volumetric specific capacitance, as well as the greatest maintenance of both the gravimetric and volumetric specific capacitance. A high volumetric energy density of 6.6 Wh L^{-1} was attained at the high-power density of 1 kW L^{-1}, which was a 35% increment compared with the nonmilled AC. The electrode densification (decreased interparticle gap) and the enhanced ion-transportation within the AC pores, which were attributed to the pulverization, were responsible for those excellent performances. It was also shown that excessive milling could degrade the EDLC performances because of the lowered micro- and meso-porosity and the excessive electrode densification to restrict the ion-transportation within the pores.

Keywords: activated carbon; ball milling; electric double-layer capacitor; supercapacitor; electrode; specific capacitance; energy density; power density

1. Introduction

Electric double-layer capacitors (EDLCs) are energy storage devices that have been applied to various energy storage sectors such as automobiles, power leveling for renewable energies, and portable electronic devices [1–3]. The advantages of EDLCs are high-power density and long cycle-life, when they were compared to secondary batteries. The charge–discharge processes with high-power density are based on a principle of physical adsorption and desorption of electrolytic ions onto the electrode, so as to allow the formation and release of the electric double-layer. Activated carbons (ACs) are mainly employed as the active materials of EDLC electrodes due to their high specific surface area, excellent cost-performance, and large-scale productivity [4]. The specific surface area of the active materials is closely related to the formation area of the electric double-layer. Highly porous carbons, whose specific surface areas are >3000 m^2 g^{-1}, have been produced from various precursors and via novel chemical activation techniques [5–9]. ACs with a high specific surface area are very effective in increasing the gravimetric energy and power density of EDLC electrodes. ACs with a specific surface area of >3000 m^2 g^{-1} are now available on the market and can be employed as the electrode active material. Those provided a very high gravimetric specific capacitance of 209–575 F g^{-1} under the use of aqueous electrolytes [6,10,11] and that of 144–338 F g^{-1} under the use of nonaqueous electrolytes [5,12,13]. Microporous ACs with a narrow width of <2 nm can produce a wide electric double-layer and thus give the electrode a high specific capacitance [14–17]. In addition to the specific surface area, the pore size distribution of ACs is also an important factor determining the capacitive and resistive behaviors of EDLCs.

The EDLC electrodes have been chiefly prepared in two ways. One is to coat a slurry comprising active material, conductive agent, and binder onto an Al foil used as the current collector [18–20]. The other is to knead the mixture composed of active materials, conductive agent, and binder, and then to mold it into a sheet. The sheet is pressed and attached with the mesh-like current collector or is adhered with a foil-like current collector [21–23]. The particle size of AC used as the active material can be industrially recommended to be in the range of 4–8 μm [24]. A correlation between the particle size of the AC active material and the charge–discharge performance of EDLC has been explored. It was demonstrated that, for the EDLC electrodes, an increase in the AC particle size led to a decrease in the specific capacitance and an increase in the equivalent series resistance (ESR) [19]. The ESR of the electrode is chiefly attributed to the electrolyte, the contact, and the intrinsic resistance. The contact resistance is related to the contact degree between the AC particles, and that between the particles and the current collector. The intrinsic resistance is dependent on the conductivity of AC particles, which have a relation with their graphitization degree and their porosity. The specific surface area of the used AC and the exposure degree of pore-walls that allow access of the electrolytic ions so as to form an electric double-layer, are decisive factors of the specific capacitance of EDLC electrode. The exposure degree of pore-walls is affected by the affinity of AC particles with the binder, and the level of binder added in the electrode. The AC particle size and its size uniformity have been known to participate in the ESR [21–23,25]. The role of AC particle size on the diffusivity of ions in the EDLC electrodes has been also investigated, suggesting that smaller AC particles facilitated ion-transportation in the EDLC electrodes [26].

In order to obtain the desired particle size, or desired particle size distribution of ACs, a ball milling technique has been well employed. It has been shown that the porous structure and surface morphology of ACs can vary during the balling milling process [27–34]. The electrodes using nanoscale carbide-derived AC powders displayed an excellent electrical contact between the particles across the electrode and facilitated ion-movement within the pores, maintaining the specific capacitance of the electrode even at high current densities [35]. The AC processed by the prolonged ball milling decreased the specific capacitance due to particle agglomeration [34]. The presence of submicron sized particles in discrete AC particle clusters led to performance and cycle stability degradation of the electrodes [21]. In the abovementioned studies, the specific surface areas of the used ACs were in the range of 880–2200 m^2 g^{-1}. Increasing the gravimetric energy and power densities of the EDLC electrode requires highly porous ACs. However, there are few reports on the optimization of the particle size of ACs with a very high specific surface area (~3000 m^2 g^{-1}), which are intended for use in the fabrication of high-performance EDLC electrodes. The effects of the pulverization degree of ACs with such high specific surface area on their EDLC performances have not been explored so far.

In the present study, the EDLC performances of ACs with specific surface areas of around 3000 m^2 g^{-1} were examined under the use of a nonaqueous electrolyte, which has been a mainstream electrolyte for high-performance commercial EDLCs. We investigated the variation in the textural properties and the particle morphology of the highly porous ACs during planetary ball milling. The charge–discharge performances (specific capacitance, rate and cyclic stabilities, and Ragone plot, from the viewpoints of both the gravimetric and volumetric performances) of EDLCs using the AC active materials of various particle sizes were evaluated. In order to achieve the greatest gravimetric or volumetric EDLC performance, the ball milling process was optimized for the highly porous ACs.

2. Materials and Methods

2.1. Ball Milling and Materials Characterization

Maxsorb (Kansai Coke and Chemicals Co., Ltd., Amagasaki, Japan), which was manufactured from petroleum coke by means of KOH activation [36], was used as the AC active material. For the ball milling process, planetary ball milling equipment (P-6, Fritch Japan Co., Ltd., Yokohama, Japan), a zirconia bowl with an 80 mL-milling space, and two types of zirconia balls (ϕ10.0 or ϕ19.5 mm) were employed. In air atmosphere, one gram of the sample AC was pulverized for 10, 90, and 120 min using the five ϕ19.5 mm balls at a rotation speed of 400 rpm. The ACs, pulverized for 10, 90, and 120 min, were termed AC10, AC90, and AC120, respectively. The AC milled for 10 min using the thirteen ϕ10.0 mm balls (ϕ19.5 mm balls were not mixed) at the similar rotation speed was also prepared to realize the particle size between those of AC0 and AC10, which was termed ACS10. The AC sample which was not milled was termed AC0.

Crystalline structures of the ACs were analyzed using an X-ray diffractometer (RINT-2020V, Rigaku Corp., Askishima, Japan) with Cu-Kα radiation (wavelength: 0.15418 nm), providing X-ray diffraction (XRD) patterns of the ACs. The disorder degree of graphene structures of the ACs was also evaluated using a microscopic Raman spectrometer (LabRAM HR Evolution, Horiba Ltd., Kyoto, Japan). The Raman spectra were acquired using a laser of 633 nm wave length. The degree of graphene disorder of ACs was quantified by the peak intensity ratio of the G-band (ca. 1580 cm^{-1}) and D-band (ca. 1360 cm^{-1}), which was defined as I_d/I_g. The D-band and G-band corresponded to the in-plane vibrations of sp^2 bonded carbon structures with structural defects and the in-plane vibrations of sp^2 bonded graphene carbon structures, respectively [37–39]. Thus, I_d/I_g is an indication of the disorder degree of graphene sheets.

The N$_2$ adsorption–desorption isotherms of the ACs were measured using a gas adsorption analyzer (Autosorb-3B, Quantachrome Instruments Inc., Boynton Beach, FL, USA) at −196 °C. Approximately 30 mg of powdered AC was degassed under a vacuum at 200 °C for >8 h prior to the isotherm measurement. The Brunauer–Emmett–Teller (BET) theory was used to calculate the specific surface area (S_{BET}) of the AC samples using the adsorption isotherm at a relative pressure of 0.05–0.10. The total pore volume (V_{total}) of the sample was determined by measuring the volume of nitrogen absorbed at the relative pressure of 0.99. The quenched solid density functional theory (QSDFT) was used to obtain the pore size distribution [40,41], with aid from the proprietary software (ASiQwin, version 1.11, Quantachrome Instruments Inc., Boynton Beach, FL, USA). Based on the pore size distributions, the volumes of the micro-pores (V_{micro}) and meso-pores (V_{meso}) were calculated.

The particle size distribution of the produced AC powder was analyzed using a laser diffraction particle size analyzer (SALD-200V, Shimadzu Corp., Kyoto, Japan). The average particle diameter and the cumulative 25%, 50% (median value), and 75% particle diameters, were acquired and were defined as D_{ave}, D_{25}, D_{50}, and D_{75}, respectively.

2.2. Electrochemical Characterization

AC as the active material, acetylene black (Denka Black, Denka Kagaku Co., Ltd., Tokyo, Japan) as the conductive agent, and polytetrafluoroethylene (Polyflon D210-C, Daikin Industries, Ltd., Osaka, Japan) as the binder were mixed in a mass ratio of 8:1:1 by a mortar and pestle, with added ethanol. The mixture was pressed into a sheet, then it was punched out into disks of ϕ12 mm. The mass of AC in the electrode and the electrode thickness were measured, providing the AC bulk density. The morphology of electrodes was microscopically observed using a scanning electron microscope (VE-8800, Keyence Corp., Osaka, Japan).

The meso- and macro-porosity, and interparticle porosity of the prepared EDLC electrodes were evaluated using a mercury porosimeter (PASCAL 140 and 240, Thermo Fisher Scientific KK, Tokyo, Japan). For each AC sample, four pieces of the same type of electrode, which were previously degassed in a vacuum at 140 °C for >6 h were subjected to a mercury intrusion and extrusion sequence. This sequence was performed in the low-pressure range (0.2–400 kPa) using PASCAL 140 and in the high-pressure range (up to 200 MPa) using PASCAL 240, providing the mercury intrusion volume at different equilibrated mercury pressures. The cumulative pore volumes of EDLC electrodes were calculated using the Washburn equation [42] with aid from a software installed in the above mercury porosimeter, where the shape of pores were assumed to be a cylinder.

The disk was pressed onto a ϕ15 mm Al mesh at a pressure of 0.5 MPa using a perpendicular press. The disk attached with an Al mesh was employed as the electrode. The electrodes were dried under a vacuum at 140 °C for >6 h prior to the cell assembly. A two-electrode cell made from Al (HS cell, Hohsen Corp., Osaka, Japan) was assembled in a glove box (GBJF080R, Glovebox Japan Inc., Inagi, Japan) filled with argon gas. The cell was constituted from two identical electrodes, a ϕ23 mm paper-based separator (TF4050, Nippon Kodoshi Corp., Kochi, Japan), and 1 mL electrolyte. The electrolyte was tetraethylammonium tetrafluoroborate at 1 mol L^{-1} dispersed in propylene carbonate (TEA·BF_4/PC, Kishida Chemical Co., Ltd., Osaka, Japan).

The current passing through the electrodes under voltage application across the cell terminals was evaluated by cyclic voltammetry (CV). CV was performed at the scan rates of 1, 10, and 100 mV s^{-1} using an electrochemical measurement system (HZ5000, Hokuto Denko Corp., Tokyo, Japan) at the cell voltage range of 0–2.5 V. The specific capacitance of the AC during the CV measurement (C_{CV}) was calculated from Equation (1).

$$C_{CV} \ (\text{F g}^{-1}) = \frac{4I_{CV}}{mV_S} \tag{1}$$

where m (g) is the total mass of AC incorporated in both the positive and negative electrodes, I_{CV} (A) is the current measured at different applied voltages, and V_S (V s^{-1}) is the voltage scan rate.

A battery charge–discharge system (HJ1005SD8, Hokuto Denko Corp., Tokyo, Japan) was used to perform the galvanostatic charge–discharge (GCD) tests for the EDLC cells, during which the cell voltage was increased to 2.5 V for the charge process and was decreased to 0 V for the discharge process at different current densities of 0.1–100 mA cm^{-2}. Table 1 shows the detailed conditions in the GCD rate tests. The charge–discharge performances at different current densities were evaluated at the specified cycle numbers. The gravimetric specific capacitance of the AC in the GCD tests (C_{GCD}) was calculated using Equation (2).

$$C_{GCD} \ (\text{F g}^{-1}) = \frac{4Q}{mV'} \tag{2}$$

where Q (C) is the charge released during the discharge process and V' (V) is the maximum cell voltage (2.5 V) subtracted by the IR drop observed at the switching from charge to discharge processes. The IR drop is the voltage drop caused by the internal resistance, which can be represented by ESR. The relationship between the current density and the ESR at the switching of charge to discharge was also obtained for all types of ACs. The volumetric specific capacitance (F cm^{-3}) was also calculated based on the AC bulk density in the electrode. The charge or discharge specific capacity of the AC was also defined to be a time-integral of the current divided by m, providing the specific capacity–cell voltage profile of the AC. The Ragone plot was created by calculating the energy density of the electrode, E (Wh kg^{-1}), and power density of the electrode, P (W kg^{-1}), from the GCD test results using Equations (3) and (4), respectively.

$$E \ (\text{Wh kg}^{-1}) = \frac{1000W}{m} \tag{3}$$

$$P \ (\text{W kg}^{-1}) = \frac{1000V'I}{m} \tag{4}$$

where W (Wh) is the energy released from the cell during the discharge process, and I (A) is the discharge current. Using the AC bulk density in the electrode, the volumetric energy density and volumetric power density, in respective units of Wh L^{-1} and W L^{-1}, were also calculated.

Table 1. Conditions for the galvanostatic charge–discharge (GCD) rate tests.

Sequence	Current Density [1] (mA cm^{-2})	Number of Cycles	Cycle Selected for the Performance Evaluation
1	0.1	2	Second
2	0.2	2	Second
3	0.5	2	Second
4	1.0	5	Third
5	2.0	5	Third
6	5.0	5	Third
7	10	11	Sixth
8	20	11	Sixth
9	50	11	Sixth
10	100	25	Thirteenth

[1] Electrode area is 1.13 cm^2 (ϕ12 mm).

For the evaluation of electrode cycle stability, following the GCD rate test, the EDLC cell was charged and discharged 2000 times at the constant current density of 10 mA cm^{-2} under the similar cell voltage range. The gravimetric specific capacitance of the electrode was measured as a function of the number of cycles. All measurements and analyses were performed at 25 °C.

3. Results and Discussion

3.1. Material Properties of the Milled ACs

The ACs with different particle sizes were produced by changing the ball milling process. XRD patterns and Raman spectra for the milled ACs were obtained and are shown in Figure 1. In the XRD patterns, all ACs exhibited two gentle humps at the 2θ values of ca. 23 and 43°. These patterns are characteristic of turbostratic carbons, meaning that these lack correspondence between graphene planes [43]. Only a tiny peak (marked by an inverted triangle) was observed in the XRD pattern of AC120 at 30°, which was confirmed to be caused by the debris of zirconia balls and bowl used for the AC milling [44]. It was clearly indicated that carbon structures of all the ACs were turbostratic, and the milling process had little impact on those carbon structures. The Raman spectra of all the ACs displayed the D-band and G-band peaks resulting from carbonaceous structures. The I_d/I_g values of AC0, AC10, ACS10, and AC90 were similar; while that of AC120 was slightly lower than the others. The lowering of I_d/I_g indicates that AC120 alleviated the disorder of graphene sheets. From both the results of X-ray diffractometry and Raman spectroscopy, it was shown that, when the ball milling time was limited to within 120 min, the structural changes of the AC during the ball milling were minor.

Figure 2 shows the nitrogen adsorption–desorption isotherms at the temperature of −196 °C. All of the nitrogen adsorption isotherms exhibited a slightly hysteretic behavior at the relative pressure of 0.2–0.8. These can be categorized into an IUPAC type I (b) isotherm, suggesting pore size distributions over a broader range including wider micropores and possibly narrow meso-pores (<~2.5 nm) [45]. It was found that, over the entire range of relative pressures, the adsorption quantity decreased with the milling time. All of the ACs allowed a development of pores that were 0.6–1.0 and 1.2–3.0 nm. The peaks of pore width for the ACs subjected to the ball milling shifted to the smaller side. Table 2 shows the textural properties of the ACs based on the above isotherms and pore size distributions. All of the textural parameters decreased with the milling time. The S_{BET} and V_{total} of the nonmilled AC (AC0) were measured to be 3198 m^2 g^{-1} and 1.78 cm^3 g^{-1}, respectively. The S_{BET} of AC90 and AC120 were 2755 and 2448 m^2 g^{-1}, respectively, indicating that the ball milling for 90 or 120 min induced surface area decrements of 13.9% and 23.5%, respectively. The ratio of V_{micro} to V_{tatal} and that of V_{meso} to V_{tatal} were ~70% and ~20%, respectively, which were similar in all ACs.

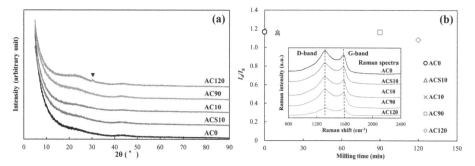

Figure 1. XRD patterns and a relationship between the milling time and I_d/I_g value for the milled activated carbons (ACs). (**a**) XRD patterns, (**b**) I_d/I_g value with their Raman spectra.

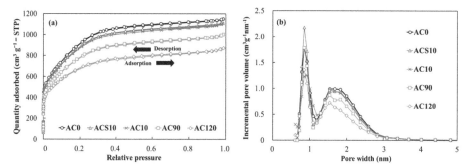

Figure 2. Nitrogen adsorption–desorption isotherms of the milled ACs at −196 °C and their pore size distributions. (**a**) Nitrogen adsorption–desorption isotherms; (**b**) pore size distributions.

Table 2. Textural properties of the milled ACs.

Sample Name	S_{BET} (m^2 g^{-1})	V_{total} (cm^3 g^{-1})	V_{micro} (cm^3 g^{-1})	V_{meso} (cm^3 g^{-1})
AC0	3198	1.78	1.27	0.38
ACS10	3073	1.73	1.23	0.37
AC10	3041	1.72	1.22	0.37
AC90	2755	1.56	1.09	0.34
AC120	2448	1.36	0.96	0.28

Figure 3 shows the surface morphology of the electrodes using the milled ACs, and their particle size distributions. With increasing the milling time, the size of AC particles decreased and the gap between the AC particles at the electrode surfaces decreased. It was observed that dense surface structures were produced for the electrodes comprising ACs that were subjected to a long milling time (AC90 and AC120). The particle size distributions suggested that the mode (most frequent) particle size of AC decreased with the milling time. AC0 was composed of course particles and allowed a dominant existence of particles at >50 μm in diameter, while AC120 comprised only fine particles with diameters of <6 μm. It should be noticed that AC90 exhibited two peaks on its particle size distribution, which were attributed to milled fine particles at <8 μm and residual large particles at 20–30 μm. Particle properties of the milled ACs calculated based on the above distributions are shown in Table 3. AC0 and ACS10, which were composed of large size particles, did not show coincident D_{ave} and D_{50}. It was also confirmed that 75% of their particles were within $2 \times D_{50}$ for AC10, AC90, and AC120, indicating that those were sufficiently pulverized rather than AC0 and ACS10. The ACs pulverized for 150 min using the five ϕ19.5 mm balls were also prepared and its D_{ave} was measured to be 3.4 μm. It was suggested that excessive milling induced a particle agglomeration and the milling for 120 min attained the greatest pulverization.

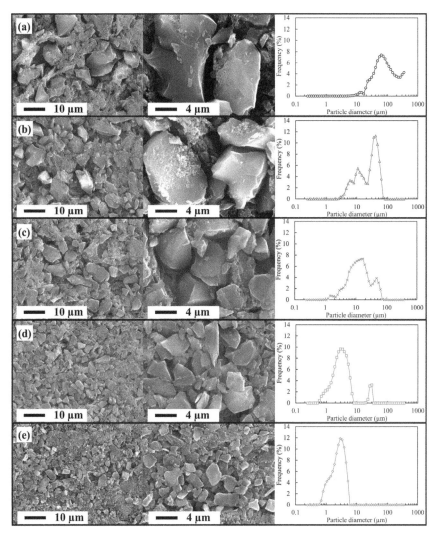

Figure 3. Surface morphology of the electric double-layer capacitor (EDLC) electrodes using the milled ACs and their particle size distributions. (**a**) AC0; (**b**) ACS10; (**c**) AC10; (**d**) AC90, (**e**) AC120.

Table 3. Particle properties of the milled ACs.

Sample Name	D_{ave} (µm)	D_{25} (µm)	D_{50} (µm)	D_{75} (µm)
AC0	74.9	42.5	70.7	135.1
ACS10	20.0	10.6	26.1	37.5
AC10	11.6	6.8	11.7	19.7
AC90	3.0	1.9	2.8	4.2
AC120	2.2	1.6	2.4	3.3

3.2. Electrochemical Properties of the Milled ACs

Details of the fabricated electrodes used for EDLC cells are shown in Table 4. The electrodes were ϕ12 mm disks incorporating the milled ACs of 10–13 mg, and their thickness ranged from 0.29 to 0.32 mm. The bulk density of AC in the electrode was dependent on the milling time. The higher AC bulk density was obtainable from the ACs pulverized for a longer time. The markedly high AC bulk density (0.39 g cm^{-3}) was apparent for the electrode using AC120.

Table 4. Details of the fabricated electrodes (ϕ12 mm).

Used AC	Mass of AC in the Electrode (mg)	Thickness of the Electrode (mm)	AC Bulk Density in the Electrode (g cm^{-3})
AC0	11.7	0.32	0.32
ACS10	10.8	0.30	0.32
AC10	10.6	0.30	0.31
AC90	12.1	0.31	0.34
AC120	12.9	0.29	0.39

The meso- and macro-porosity, and interparticle porosity of the prepared EDLC electrodes were evaluated by means of mercury porosimetry. Figure 4 shows the cumulative pore volumes of the electrodes using the milled ACs. With decreasing the pore diameter, in other words, with increasing the mercury intrusion pressure, the cumulative volume of mercury intruded into the EDLC electrodes increased. All the electrodes allowed a linear increase in the cumulative intruded volume up to ~0.4 cm^3 g^{-1}. Above ~0.4 cm^3 g^{-1}, the cumulative intruded volume steeply increased with decreasing the pore diameter. Judging from the surface morphology of the EDLC electrodes, shown in Figure 3, the pore diameter corresponding to the end of the linear volume increase coincided with the interparticle distance (gap between AC particles), for instance, 1.5 µm for AC10, 0.7 µm for AC90, and 0.3 µm for AC120. Thus, it is reasonable to interpret that the variations in the cumulative intruded volume above 0.4 cm^3 g^{-1} indicates the meso- and macro-porosity of the electrodes, which were produced within AC particles. It was observed that the electrodes using ACS10 and AC10 allowed a greater development of macro-pores, while that using AC90 possessed a largest volume of meso-pores of 7–50 nm. The electrodes using AC0 and AC120 exhibited the poorest macro- and meso-porosity. The electrodes using ACS10 and AC10 had a similar meso- and macro-porosity, although their average particle sizes had a twofold difference.

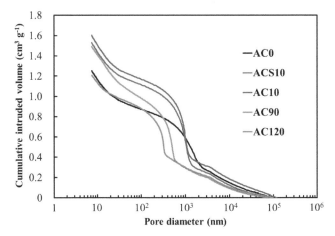

Figure 4. Cumulative pore volumes of the disk EDLC electrodes using the milled ACs, obtained by mercury intrusion porosimetry.

The cyclic voltammograms converted from the current response during the CV at the scan rates of 1, 10, and 100 mV s^{-1} are shown in Figure 5, providing the specific capacitance at the varying voltage cell. The symmetrical curves appeared in all the ACs at the scan rate of 1 mV s^{-1}, meaning that an electric double-layer without a redox reaction was soundly formed on the electrode. At the scan rate of 1 mV s^{-1}, AC0 showed the highest gravimetric specific capacitance, and the gravimetric specific capacitance of the ACs decreased with the milling time. As the voltage scan rate increased, the CV curves allowed a depressive distortion. The curve distortion of AC90 was found to be most suppressed at the scan rates of 10 and 100 mV s^{-1}, indicating that the lowest internal resistance was produced in the AC90 electrode. It should be noticed that, at all the scan rates, the gravimetric specific capacitance of AC120 was lowest in all the ACs. The highest gravimetric specific capacitance did not appear for the nonmilled and the most milled ACs at the increased scan rates, suggesting that the degree of milling had a definite influence on the charge–discharge rate performance of AC.

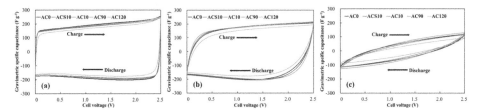

Figure 5. Cyclic voltammetry (CV) curves of EDLC cells using the milled ACs at the scan rates of (**a**) 1, (**b**) 10, and (**c**) 100 mV s^{-1}.

Figure 6 shows the specific capacity–cell voltage profiles of the ACs under different current densities during the GCD rate test. The IR drops were observed at the start of charging and discharging. The IR drops appeared distinctly above the current density of >10 mA cm^{-2} in all of the samples. AC120 allowed the largest IR drop, while the other ACs allowed similar levels of IR drops, indicating that a noticeable ESR was produced within the AC120 electrode. At the lower current density of <1 mA cm^{-2}, where the IR drop was negligible, a longer milling time led to a decrease in the specific capacity at the end of the discharge process. Representative ESR values of the milled ACs at the switching of charge to discharge during the rate GCD tests were calculated and are shown in Table 5. The ESRs of the milled ACs tended to decrease with the current density. The ESR of AC120 was found to be considerably high, while those of the other ACs were comparable regardless of the current density. The comparable ESR values mean that the charge-storage or charge-release performance of ACs was mainly governed by their dynamics of double-layer formation and release, and was hardly related to the electrolyte, contact, and intrinsic resistances.

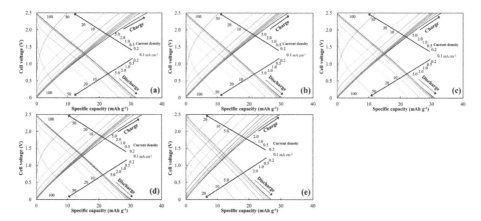

Figure 6. Specific capacity of milled ACs vs. cell voltage in the GCD rate tests at different current densities. The specific capacity is a time-integral of the current divided by the total mass of AC active material both in the positive and negative electrodes. (**a**) AC0; (**b**) ACS10; (**c**) AC10; (**d**) AC90; (**e**) AC120.

Table 5. Equivalent series resistances (ESRs) of the milled ACs at the switching of charge to discharge during the GCD rate tests. Unit: Ω.

Current Density (mA cm^{-2})	AC0	ACS10	AC10	AC90	AC120
0.1	46.4	47.8	46.1	46.7	141.7
1	19.1	15.3	19.6	15.2	108.0
5	16.4	16.2	16.8	12.7	96.8
10	13.8	14.3	14.1	11.9	73.9
50	13.3	14.0	13.7	12.5	NM [1]
100	12.5	13.8	13.5	12.6	NM [1]

[1] Not measurable.

The gravimetric and volumetric specific capacitances, and the gravimetric and volumetric Ragone plots for all the ACs are shown in Figure 7. The gravimetric specific capacitances of all the ACs were maintained up to the current density of 5 mA cm^{-2}. The nonmilled AC (AC0) displayed the highest gravimetric specific capacitance of 183 F g^{-1}, while the AC subjected to the longest milling (AC120) had the lowest value (<160 F g^{-1}). At the higher current density (>10 mA cm^{-2}), the ACs allowed a decline of the gravimetric specific capacitance, where the degree was dependent on the milling time. AC10 and AC90, which had middle levels of D_{ave}, could retain a high gravimetric specific capacitance. Under a very high current density (>20 mA cm^{-2}), the gravimetric specific capacitance of AC120 was measured to be negligible. From the gravimetric Ragone plots, the highest gravimetric energy density of 39.3 Wh kg^{-1} was observed on AC0 at the gravimetric power density of 12.0 W kg^{-1}. Except for AC120, very little difference in gravimetric energy density among the type of ACs was observed at the lower gravimetric power density (<1000 W kg^{-1}). Under the very high gravimetric power density (>2000 W kg^{-1}), AC90 displayed the highest gravimetric energy density of all the ACs.

The volumetric specific capacitances of all the ACs were different from the gravimetric ones. At the lower current density (<10 mA cm^{-2}), AC120 showed the highest volumetric specific capacitance of 61.7 F cm^{-3}. Sufficiently milled ACs (AC90 and AC120) could also attain a similar level of the high volumetric specific capacitance. However, under the higher current density (>10 mA cm^{-2}),

AC0 and AC120 allowed the volumetric specific capacitance to lower. Only AC90 could largely retain the volumetric specific capacitance. The volumetric Ragone plots indicated that all the ACs showed a comparable volumetric energy density at the lower volumetric power density (<100 Wh L^{-1}). Among them, the highest volumetric energy density of 13.0 Wh L^{-1} was observed for AC120 at the volumetric power density of 4.3 W L^{-1}. AC90 had the highest ability to keep the volumetric energy density, even under the higher volumetric power density (>100 W L^{-1}). The appropriate milling process enabled us to enhance the volumetric energy density from 4.9 (AC0) to 6.6 Wh L^{-1} (AC90) even under the high volumetric power density of 1 kW L^{-1}, while the most milled AC (AC120) did not operate at 1 kW L^{-1}. It is true that the differences between the particle properties of AC90 and AC120 were not as large (D_{ave} = 3.0 μm for AC90 and D_{ave} = 2.2 μm for AC120). The results regarding the GCD rate tests revealed that the ball milling process had a large influence on the volumetric performance, rather than on the gravimetric performance. It was also revealed that excessive milling of the AC active material can degrade the high-power performance of the EDLC cell.

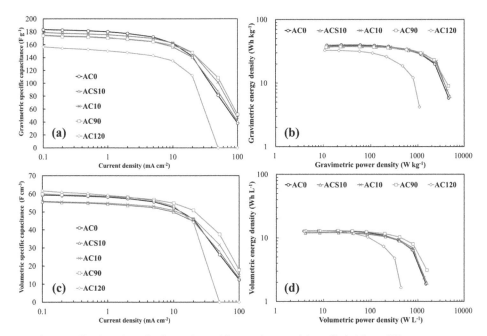

Figure 7. Gravimetric and volumetric specific capacitances of the milled ACs at different current densities and their Ragone plots. (**a**) Gravimetric specific capacitance; (**b**) gravimetric Ragone plot; (**c**) volumetric specific capacitance; (**d**) volumetric Ragone plot.

Figure 8 shows the cyclic stability of the specific capacitance of the milled ACs at the current density of 10 mA cm^{-2} under the cell voltage range of 0–2.5 V. The specific capacitance retentions of AC0, ACS10, AC10, AC90, and AC120 were 96%, 93%, 96%, 95%, and 93%, respectively, which guaranteed the initial cycling stability of ACs. Although their long-term stability could not be verified, the effect of the milling process on the cyclic charge–discharge performance was shown to be minor.

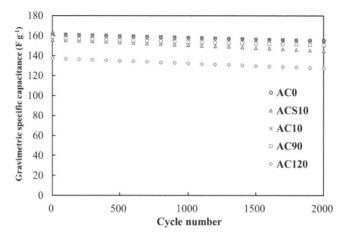

Figure 8. Cyclic stability of the specific capacitance of the milled ACs at the current density of 10 mA cm^{-2} under the cell voltage range of 0–2.5 V.

3.3. Material and Electrochemical Properties of the ACs with Different Particle Sizes

The petroleum coke-based AC with a very high specific surface area (S_{BET} = 3198 m^2 g^{-1}) and with a number of course particles (D_{ave} = 74.9 µm) was pulverized to find its greatest gravimetric or volumetric EDLC performance. The pulverization was executed by means of planetary ball milling under different milling times and different types of zirconia balls. Using five ϕ19.5 mm zirconia balls at a rotation speed of 400 rpm, finely and uniformly pulverized ACs were produced; D_{ave} = 11.6 µm for 10 min (AC10), 3.0 µm for 90 min (AC90), and 2.2 µm for 120 min (AC120). Instead of gaining fine and uniform particles, the AC allowed a decrease in S_{BET} because of the cumulative collisions between the balls and the particles. X-ray diffractometry and Raman spectroscopy revealed that structural changes of the AC hardly occurred during the ball milling. Despite the difference in D_{ave} between the AC90 and AC120 being slight, AC120 displayed a much lower S_{BET} (2448 m^2 g^{-1}) than did AC90 (2775 m^2 g^{-1}). The pore size distribution analysis verified that pores within the AC degenerated during the ball milling. A trade-off between the pulverization and the porosity maintenance was apparent up to the milling time of 90 min. However, this relation was invalid for the longer milling time (120 min). Under the present ball milling mechanism, the pulverization of particles became saturated, while the pore degeneration subsequently proceeded with the milling time.

The performances of milled ACs as the active materials of EDLC electrodes was explored by means of GCD rate tests. AC0 and AC120 exhibited the maximum gravimetric and volumetric specific capacitance, respectively. However, at the higher current density (>10 mA cm^{-2}), both the ACs allowed noticeable decreases in the specific capacitance. The greatest maintenance of both the gravimetric and volumetric specific capacitance, as well as the high volumetric specific capacitance, which was comparable to that of AC120, was observed on AC90. At the high-power density of 1 kW L^{-1}, the volumetric energy density of the electrode increased from 4.9 (AC0) to 6.6 Wh L^{-1} (AC90, 35% increment) by the milling process optimization. It was advocated that the smaller size of AC particles was responsible for the decrease in the diffusion resistance of ion-transport within the particles [26]. Even if AC90 allowed the decrease in surface area to form an electric double-layer because of the long-term milling, the electrode densification and the enhanced ion-transportation within the AC pores led to the excellent rate performances. AC120, which was most finely and uniformly pulverized, displayed the poorest rate performances out of all the ACs. The difference in statistic parameters of D_{ave} or D_{50} between AC90 and AC120 was slight. Based on the percolation theory for carbon particles in an insulating media, the particle size can have an influence on the electric conductivity of the composite [46]. However, judging from the slight difference in D_{ave} between AC90 and AC120, and the low fraction of insulating polytetrafluoroethylene binder in the electrode (10 mass%), the greater rate performance of AC90 was not explainable from a percolation theory. It was revealed that AC90 included milled fine particles and residual large particles at 20–30 μm, while AC120 comprised only fine particles of <6 μm. The much higher electrode bulk density for AC120 (0.39 g cm^{-3}) was clearly attributed to the more finely and uniformly pulverized particles. For the highly filled EDLC electrodes, a wide interparticle gap, which is necessary to assist the impregnation of electrolyte into discrete AC particles, and thus into micro- and meso-pores within the ACs [23,25], was not realized. It was also demonstrated that the electrode using AC90 attained the greatest meso-porosity of 7–50 nm, which could offer fluent ion-transportation connected to the higher rate performance.

The results obtained here indicated that, for ACs with a very high specific surface area (~3000 m^2 g^{-1}), particle pulverization was still effective for enhancing the volumetric EDLC performances and the stability against the high current or power density. However, excessive pulverization can rather degrade the EDLC performances of ACs because of excessive electrode densification, which restricts the ion-transportation within pores. In addition to the statistic particle size parameters (D_{ave} or D_{50}), the particle size distribution of AC was also found to be an important factor to determine the EDLC performances.

4. Conclusions

In the present study, the ball milling process was optimized for AC with a very high surface area and course particles, as the active material of EDLC cells using a nonaqueous electrolyte, so as to attain the greatest electrochemical performance. The carbonaceous structure, micro- and meso-porosity, and particle size distribution of the AC were evaluated as a time function of the planetary ball milling. The EDLC electrodes comprising the ACs milled under different conditions were evaluated in the GCD rate and cycle tests, providing their gravimetric and volumetric specific capacitances, Ragone plots, and cyclic charge–discharge stability. Meso- and macro-porosity, and interparticle porosity of the EDLC electrodes produced from the milled ACs were also evaluated.

With the development of pulverization, the AC allowed the decrease in S_{BET}, which was closely related to the area of double-layer formation. The AC milled at 400 rpm for 90 min (AC90) allowed a decrease in S_{BET} from 3198 to 2775 m^2 g^{-1}, and thereby a decrease in the gravimetric specific capacitance at the lower current density of <10 mA cm^{-2}. However, it exhibited the greatest maintenance of both the gravimetric and volumetric specific capacitance at the higher current density of >10 mA cm^{-2}. Even at the high volumetric power density of 1 kW L^{-1}, it showed a noticeable volumetric energy density of 6.6 Wh L^{-1}, which was 35% higher than that of the untreated AC. A combination of milled fine particles at <8 μm and residual large particles at 20–30 μm led to the electrode densification (decreased interparticle gap), and the enhanced ion-transportation within the AC pores.

It was also revealed that excessive pulverization did not enhance either the gravimetric or the volumetric performances of the AC, in particular at the higher power density. A much higher electrode bulk density (0.39 g cm^{-3}) was attained by the AC milled for 120 min (AC120), which comprised only fine particles of <6 μm. However, the lowered micro- and meso-porosity and the excessive electrode densification restricted the ion-transportation within the pores, leading to degradation of the high-power performance. Judging from a minor difference in the particle size parameter (D_{ave} or D_{50}) between AC90 and AC120, the particle size distribution of AC was found to be more important in tuning the EDLC performances of highly porous ACs. It was confirmed that the appropriate pulverization of AC particles had potential to enhance the high-power performance of EDLC cells, both from the gravimetric and volumetric viewpoints.

Author Contributions: T.E. designed and performed the experiments, analyzed the results, and wrote the manuscript; Y.K. performed the experiments and analyzed the results; M.T. and D.T. analyzed the results and supervised this research; S.K. designed the experiments, analyzed the results, wrote the manuscript, and administrated this research project. All authors have read and agreed to the published version of the manuscript.

Funding: This work was in part supported by JSPS KAKENHI, under grant number JP19H02121.

Acknowledgments: We would like to thank Makoto Yamaguchi of Akita University for his help with the Raman spectroscopy.

Conflicts of Interest: The authors declare no conflict of interest.

References

1. Zou, C.; Zhang, L.; Hu, X.; Wang, Z.; Wik, T.; Pecht, M. A review of fractional-order techniques applied to lithium-ion batteries, lead-acid batteries, and supercapacitors. *J. Power Sources* **2018**, *390*, 286–296. [CrossRef]
2. Sakka, M.A.; Gualous, H.; Mierlo, J.V.; Culcu, H. Thermal modeling and heat management of supercapacitor modules for vehicle applications. *J. Power Sources* **2009**, *194*, 581–587. [CrossRef]
3. Kurzweil, P.; Shamonin, M. State-of-charge monitoring by impedance spectroscopy during long-term self-discharge of supercapacitors and lithium-ion batteries. *Batteries* **2018**, *4*, 35. [CrossRef]
4. Liu, C.F.; Liu, Y.C.; Yi, T.Y.; Hu, C.C. Carbon materials for high-voltage supercapacitors. *Carbon* **2019**, *145*, 529–548. [CrossRef]
5. Mori, T.; Iwamura, S.; Ogino, I.; Mukai, S.R. Cost-effective synthesis of activated carbons with high surface areas for electrodes of non-aqueous electric double layer capacitors. *Sep. Purif. Technol.* **2019**, *214*, 174–180. [CrossRef]
6. Huang, G.; Wang, Y.; Zhang, T.; Wu, X.; Cai, J. High-performance hierarchical N-doped porous carbons from hydrothermally carbonized bamboo shoot shells for symmetric supercapacitors. *J. Taiwan Inst. Chem. Eng.* **2019**, *96*, 672–680. [CrossRef]
7. Zhu, Y.; Chen, M.; Zhang, Y.; Zhao, W.; Wang, C. A biomass-derived nitrogen-doped porous carbon for high-energy supercapacitor. *Carbon* **2018**, *140*, 404–412. [CrossRef]
8. Sevilla, M.; Ferrero, G.A.; Diez, N.; Fuertes, A.B. One-step synthesis of ultra-high surface area nanoporous carbons and their application for electrochemical energy storage. *Carbon* **2018**, *131*, 193–200. [CrossRef]
9. Gopiraman, M.; Deng, D.; Kim, B.S.; Chung, I.M.; Kim, I.S. Three-dimensional cheese-like carbon nanoarchitecture with tremendous surface area and pore construction derived from corn as superior electrode materials for supercapacitors. *Appl. Surf. Sci.* **2017**, *409*, 52–59. [CrossRef]
10. Peng, L.; Liang, Y.; Dong, H.; Hu, H.; Zhao, X.; Cai, Y.; Xiao, Y.; Liu, Y.; Zheng, M. Super-hierarchical porous carbons derived from mixed biomass wastes by a stepwise removal strategy for high-performance supercapacitors. *J. Power Sources* **2018**, *377*, 151–160. [CrossRef]
11. Pontiroli, D.; Scaravonati, S.; Magnani, G.; Fornasini, L.; Bersani, D.; Bertoni, G.; Milanese, C.; Girella, A.; Ridi, F.; Verucchi, R.; et al. Super-activated biochar from poultry litter for high-performance supercapacitors. *Microporous Mesoporous Mater.* **2019**, *285*, 161–169. [CrossRef]
12. Zou, Z.; Liu, T.; Jiang, C. Highly mesoporous carbon flakes derived from a tubular biomass for high power electrochemical energy storage in organic electrolyte. *Mater. Chem. Phys.* **2019**, *223*, 16–23. [CrossRef]
13. Gao, Y.; Li, L.; Jin, Y.; Wang, Y.; Yuan, C.; Wei, Y.; Chen, G.; Ge, J.; Lu, H. Porous carbon made from rice husk as electrode material for electrochemical double layer capacitor. *Appl. Energy* **2015**, *153*, 41–47. [CrossRef]
14. Raymundo-Pinero, E.; Kierzek, K.; Machnikowski, J.; Béguin, F. Relationship between the nanoporous texture of activated carbons and their capacitance properties in different electrolytes. *Carbon* **2006**, *44*, 2498–2507. [CrossRef]
15. Yang, I.; Kim, S.G.; Kwon, S.H.; Kim, M.S.; Jung, J.C. Relationships between pore size and charge transfer resistance of carbon aerogels for organic electric double-layer capacitor electrodes. *Electrochim. Acta* **2017**, *223*, 21–30. [CrossRef]
16. Dong, X.L.; Lu, A.H.; Li, W.C. Highly microporous carbons derived from a complex of glutamic acid and zinc chloride for use in supercapacitors. *J. Power Sources* **2016**, *327*, 535–542. [CrossRef]
17. Chmiola, J.; Yushin, G.; Gogotsi, Y.; Portet, C.; Simon, P.; Taberna, P.L. Anomalous increase in carbon capacitance at pore sizes less than 1 nanometer. *Science* **2006**, *313*, 1760–1763. [CrossRef] [PubMed]
18. Simon, P.; Burke, A. Nanostructured carbons: Double-layer capacitance and more. *Electrochem. Soc. Interface* **2008**, *17*, 38–43.
19. Yoshida, A.; Nonaka, S.; Aoki, I.; Nishino, A. Electric double-layer capacitors with sheet-type polarizable electrodes and application of the capacitors. *J. Power Sources* **1996**, *60*, 213–218. [CrossRef]
20. Pandolfo, A.G.; Wilison, G.J.; Huynh, T.D.; Hollenkamp, A.F. The influence of conductive additives and inter-particle voids in carbon EDLC electrodes. *Fuel Cells* **2010**, *5*, 856–864. [CrossRef]

21. Rennine, A.J.R.; Martins, V.L.; Smith, R.M.; Hall, P.J. Influence of particle size distribution on the performance of ionic liquid-based electrochemical double layer capacitors. *Sci. Rep.* **2016**, *6*, 22062. [CrossRef] [PubMed]
22. Dyatkin, B.; Gogotsi, O.; Malinovskiy, B.; Zozulya, Y.; Simon, P.; Gogotsi, Y. High capacitance of coarse-grained carbide derived carbon electrodes. *J. Power Sources* **2016**, *306*, 32–41. [CrossRef]
23. Kado, Y.; Imoto, K.; Soneda, Y.; Yoshizawa, N. Correlation between the pore structure and electrode density of MgO-templated carbons for electric double layer capacitor applications. *J. Power Sources* **2016**, *305*, 128–133. [CrossRef]
24. Azaïs, P. *Manufacturing of Industrial Supercapacitor. Supercapacitor Materials, Systems, and Applications*; Béguin, F., Frackowiak, E., Eds.; Wiley-VCH Verlag GmbH & Co. KGaA: Weinheim, Germany, 2013; p. 320.
25. Kado, Y.; Soneda, Y. Void-bearing electrodes with microporous activated carbon for electric double-layer capacitors. *J. Electroanal. Chem.* **2019**, *833*, 33–38. [CrossRef]
26. Tanaka, S.; Nakao, H.; Mukai, T.; Katayama, Y.; Miyake, Y. An experimental investigation of the ion storage/transfer behavior in an electrical double-layer capacitor by using monodisperse carbon spheres with microporous structure. *J. Phys. Chem. C* **2012**, *116*, 26791–26799. [CrossRef]
27. Nandhini, R.; Mini, P.A.; Avinash, B.; Nair, S.V.; Subramanian, K.R.V. Supercapacitor electrodes using nanoscale activated carbon from graphite by ball milling. *Mater. Lett.* **2012**, *87*, 165–168. [CrossRef]
28. Xu, J.; Zhang, R.; Wang, J.; Ge, S.; Zhou, H.; Liu, Y.; Chen, P. Effective control of the microstructure of carbide-derived carbon by ball-milling the carbide precursor. *Carbon* **2013**, *52*, 499–508. [CrossRef]
29. Welham, N.J.; Berbenni, V.; Chapman, P.G. Increased chemisorption onto activated carbon after ball-milling. *Carbon* **2002**, *40*, 2307–2315. [CrossRef]
30. Choi, W.S.; Shim, W.G.; Ryu, D.W.; Hwang, M.J.; Moon, H. Effect of ball milling on electrochemical characteristics of walnut shell-based carbon electrodes for EDLCs. *Microporous Mesoporous Mater.* **2012**, *155*, 274–280. [CrossRef]
31. Müller, B.R. Effect of particle size and surface area on the adsorption of albumin-bonded bilirubin on activated carbon. *Carbon* **2010**, *48*, 3607–3615. [CrossRef]
32. Ong, T.S.; Yang, H. Effect of atmosphere on the mechanical milling of natural graphite. *Carbon* **2000**, *38*, 2077–2085. [CrossRef]
33. Partlan, E.; Davis, K.; Ren, Y.; Aqul, O.G.; Mefford, O.T.; Karanfil, T.; Ladner, D.A. Effect of bead milling on chemical and physical characteristics of activated carbons pulverized to superfine sizes. *Water Res.* **2016**, *89*, 161–170. [CrossRef] [PubMed]
34. Macías-García, A.; Torrejón-Martín, D.; Díaz-Díez, M.Á.; Carrasco-Amador, J.P. Study of the influence of particle size of activate carbon for the manufacture of electrodes for supercapacitors. *J. Energy Storage* **2019**, *25*, 100829. [CrossRef]
35. Portet, C.; Yushin, G.; Gogotsi, Y. Effect of carbon particle size on electrochemical performance of EDLC. *J. Electrochem. Soc.* **2008**, *155*, 531–536. [CrossRef]
36. Otowa, T.; Tanibata, R.; Itoh, M. Production and adsorption characteristics of MAXSORB: High-surface-area active carbon. *Gas. Sep. Purif.* **1993**, *7*, 241–245. [CrossRef]
37. Katagiri, G.; Ishida, H.; Ishitani, A. Raman spectra of graphite edge planes. *Carbon* **1988**, *26*, 565–571. [CrossRef]
38. Eckmann, A.; Felten, A.; Mishchenko, A.; Britnell, L.; Krupke, R.; Novoselov, K.S.; Casiraghi, C. Probing the nature of defects in graphene by Raman spectroscopy. *Nano Lett.* **2012**, *12*, 3925–3930. [CrossRef]
39. Guizani, C.; Haddad, K.; Limousy, L.; Jeguirim, M. New insights on the structural evolution of biomass char upon pyrolysis as revealed by the Raman spectroscopy and elemental analysis. *Carbon* **2017**, *119*, 519–521. [CrossRef]
40. Neimark, A.V.; Lin, Y.; Ravikovitch, P.I.; Thommes, M. Quenched solid density functional theory and pore size analysis of micro-mesoporous carbons. *Carbon* **2009**, *47*, 1617–1628. [CrossRef]
41. Gor, G.Y.; Thommes, M.; Cychosz, K.A.; Neimark, A.V. Quenched solid density functional theory method for characterization of mesoporous carbons by nitrogen adsorption. *Carbon* **2012**, *50*, 1583–1590. [CrossRef]
42. Piedboeuf, M.L.C.; Léonard, A.F.; Traina, K.; Job, N. Influence of the textural parameters of resorcinol–formaldehyde dry polymers and carbon xerogels on particle sizes upon mechanical milling. *Colloids Surf. A Phys. Eng. Asp.* **2015**, *471*, 124–132. [CrossRef]
43. Li, Z.Q.; Lu, C.J.; Xia, Z.P.; Zhou, Y.; Luo, Z. X-ray diffraction patterns of graphite and turbostratic carbon. *Carbon* **2007**, *45*, 1686–1695. [CrossRef]

44. Sathyaseelan, B.; Manikandan, E.; Baskaran, I.; Senthilnathan, K.; Sivakumar, K.; Moodley, M.K.; Ladchumananandasivam, R.; Maaza, M. Studies on structural and optical properties of ZrO_2 nanopowder for opto-electronic applications. *J. Alloys Compd.* **2017**, *694*, 556–559. [CrossRef]

45. Thommes, M.; Kaneko, K.; Neimark, A.V.; Olivier, J.P.; Rodriguez-Reinoso, F.; Rouquerol, J.; Sing, K.S.W. Physisorption of gases, with special reference to the evaluation of surface area and pore size distribution (IUPAC technical report). *Pure Appl. Chem.* **2015**, *87*, 1051–1069. [CrossRef]

46. Jing, X.; Zhao, W.; Lan, L. The effect of particle size on electric conducting percolation threshold in polymer/conducting particle composites. *J. Mater. Sci. Lett.* **2000**, *19*, 377–379. [CrossRef]

Article

Synthesis of a NiMoO₄/3D-rGO Nanocomposite via Starch Medium Precipitation Method for Supercapacitor Performance

Shahrzad Arshadi Rastabi [1], Rasoul Sarraf Mamoory [1,*], Nicklas Blomquist [2], Manisha Phadatare [2] and Håkan Olin [2]

[1] Department of Materials Engineering, Tarbiat Modares University, 14115111 Tehran, Iran;
    Shahrzad.Arshadi@modares.ac.ir
[2] Department of Natural Sciences, Mid Sweden University, 85170 Sundsvall, Sweden;
    Nicklas.Blomquist@miun.se (N.B.); Manisha.Phadatare@miun.se (M.P.); Hakan.Olin@miun.se (H.O.)
*   Correspondence: rsarrafm@modares.ac.ir; Tel.: +98-912-133-4979

Received: 30 September 2019; Accepted: 9 December 2019; Published: 15 January 2020

Abstract: This paper presents research on the synergistic effects of nickel molybdate and reduced graphene oxide as a nanocomposite for further development of energy storage systems. An enhancement in the electrochemical performance of supercapacitor electrodes occurs by synthesizing highly porous structures and achieving more surface area. In this work, a chemical precipitation technique was used to synthesize the NiMoO₄/3D-rGO nanocomposite in a starch media. Starch was used to develop the porosities of the nanostructure. A temperature of 350 °C was applied to transform graphene oxide sheets to reduced graphene oxide and remove the starch to obtain the NiMoO₄/3D-rGO nanocomposite with porous structure. The X-ray diffraction pattern of the NiMoO₄ nano particles indicated a monoclinic structure. Also, the scanning electron microscope observation showed that the NiMoO₄ NPs were dispersed across the rGO sheets. The electrochemical results of the NiMoO₄/3D-rGO electrode revealed that the incorporation of rGO sheets with NiMoO₄ NPs increased the capacity of the nanocomposite. Therefore, a significant increase in the specific capacity of the electrode was observed with the NiMoO₄/3D-rGO nanocomposite ($450\,\mathrm{Cg}^{-1}$ or $900\,\mathrm{Fg}^{-1}$) when compared with bare NiMoO₄ nanoparticles ($350\,\mathrm{Cg}^{-1}$ or $700\,\mathrm{Fg}^{-1}$) at the current density of $1\,\mathrm{A\,g}^{-1}$. Our findings show that the incorporation of rGO and NiMoO₄ NP redox reactions with a porous structure can benefit the future development of supercapacitors.

Keywords: electrochemical performance; starch; porous structure; NiMoO₄/3D-rGO nanocomposite; NiMoO₄ NPs

---

1. Introduction

Green energy plays a key role in the development of modern human life and the advancement of new technologies. Electrochemical energy storage (EES) systems like batteries and supercapacitors are among the various energy storage systems that have received great attention currently. Supercapacitors, in particular, have generated great interest due to their high power density and long cycling life [1]. Better performance of the electrodes can be achieved by the development and modification of the new materials [2]. Generally, three types of electrode materials—carbon materials, metal oxide, and conducting polymers—have been used for supercapacitors [3]. However, transition metal oxides and hydroxides, such as NiO [4], Ni(OH)₂ [5–7], MnO₂ [8,9], MoO₃ [10], and Co₃O₄ [11], are mostly employed as the supercapacitors material. These types of electrode materials are low cost and naturally abundant, with significant specific capacity based on redox reactions and high electrochemical activity [12]. Lately, binary metal oxides such as NiCo₂O₄ [13], NiMoO₄ [14,15], CoMoO₄ [16], and $MnMoO_4$ [17] have attracted significant attention as good candidates for supercapacitors. Binary metal oxides present different oxidation states and significantly higher electrical conductivity than single metal oxides. Among these binary metal oxides, $NiMoO_4$ has attracted significant research interest because of the high electrical conductivity of the Mo element and the high electrochemical activity attributed to rich redox reactions of the nickel ions [18]. Most of the articles on $NiMoO_4$ has adopted the theory of a pseudocapacitive charge storage mechanism, but a few recent articles have treated the material as battery-like. There seems to be theories in contradiction in this field. However, since this study does not focus on the specific charge storage mechanism, along with the references, we have adopted the theory of pseudocapacitive behavior. In addition, $NiMoO_4$ nanoparticles (NPs) are readily available and highly stable in alkaline electrolytes [19]. Furthermore, hybridizing metal oxides with carbon materials can be an effective method to improve the electrical conductivity and the performance of the supercapacitors [20]. This encouraged us to fabricate reduced graphene oxide (rGO) and $NiMoO_4$ nanocomposite with a unique nanostructure, which will combine the advantages of both rGO with low electrical resistance and the quick redox reactions of $NiMoO_4$.

Various synthesis methods have been studied to produce the $NiMoO_4$ nanostructures, such as co-precipitation [21], sol-gel [14], microwave-assisted solvothermal [22], sonochemical [23], microwave sintering [24], and hydrothermal [25] methods. In this work, we propose a facile and an efficient method to synthesize a new three-dimensional hybrid structure, combining $NiMoO_4$ NPs and rGO. This method can produce large amounts of material with less equipment and low temperatures compared to other techniques. However, the highly porous structure is the key parameter in supercapacitors, and a comprehensive report on the synthesis of such porous $NiMoO_4$/3D-rGO nanostructures by a facile method is lacking. Therefore, starch was used in this study during the preparation of the nanocomposite by a simple precipitation method to gain porosity and surface area. The porous structure provides a rapid ion diffusion path and increases the capacity.

2. Experimental

2.1. Materials

Sodium molybdate dihydrate ($Na_2MoO_4 \cdot 2H_2O$), nickel nitrate hexahydrate ($Ni(NO_3)_2 \cdot 6H_2O$), graphene oxide (GO), and potassium hydroxide (KOH) were provided from Merck, Sundsvall, Sweden. Polyvinylidene fluoride (PVDF), dimethylformamide (DMF), and starch (($C_6H_{10}O_5)_n$) were purchased from Sigma-Aldrich, Sundsvall, Sweden. In addition, distilled water was used throughout the sample preparation.

2.2. Synthesis of $NiMoO_4$ Nanoparticles and $NiMoO_4$ NPs/rGO Nanocomposite

The $NiMoO_4$ NPs/rGO nanocomposite was produced with a chemical precipitation method. At first, a solution of starch was prepared at 80 °C (containing 1 g of starch in 20 mL of H_2O). Then, 1 mg of GO was added to the prepared starch solution. Next, 20 mL of 50-mM $Ni(NO_3)_2 \cdot 6H_2O$ and 20 mL of 50-mM $Na_2MoO_4 \cdot 2H_2O$ were prepared separately at room temperature. Following this, $Ni(NO_3)_2 \cdot 6H_2O$ was first added to the mixed solution of starch/GO, and after 10 min of stirring through a magnetic stirrer, 50-mM $Na_2MoO_4 \cdot 2H_2O$ was added to the above solution with continued stirring for 1 h at the temperature of 80 °C. The products were collected by centrifugation at 8000 rpm and distilled water was used to wash the prepared powder, and then a vacuum oven with a temperature of 60 °C was applied to dry the samples for 24 h. Finally, the as-prepared powder was annealed at the temperature of 350 °C for 2 h in an Ar atmosphere to produce the $NiMoO_4$/3D-rGO nanocomposite. The same experiment without adding GO was performed to prepare pure $NiMoO_4$ NPs.

2.3. Fabrication of $NiMoO_4$ NPs and $NiMoO_4$/3D-rGO Nanocomposite Electrodes

For the electrochemical study, the Ni foam surface was first washed repeatedly with HCl (37 wt.%) and ethanol. The electrodes were fabricated by mixing an electroactive material ($NiMoO_4$

or NiMoO$_4$/3D-rGO-80 wt.%), carbon black (10 wt.%), and PVDF (10 wt.%) in DMF as the solvent. The mixture was subjected to intense ultrasonic treatment for 10 min to form a homogeneous paste. The final mixture was mechanically pasted on the Ni foam surface (1 × 1 cm^2) and dried at 70 °C for 12 h. The active mass loading on the Ni foam surface was estimated to be 2 mg·cm^{-2} for each electrode.

2.4. Material Characterization

The prepared powder was characterized using several analytical systems. Powder X-ray diffraction (XRD, Bruker D2 Phaser, Cu K$_\alpha$ radiation, Sweden) was used to investigate the composition of the as-prepared samples. The morphology and microstructure of the powder were characterized using field emission scanning electron microscopy (SEM assisted with Energy Dispersive X-ray Analysis (EDX), TESCAN MAIA3-2016, operated at 10.0 kV, Sweden). Raman spectroscopy (Horiba Xplora plus, laser excitation at 532 nm-frequency range of 50 to 3000 cm^{-1}) and Fourier transform infrared spectrometry (FTIR, ABB Bomem System-KBr method, Sweden-frequency range of 400 to 4000 cm^{-1}) were also used for the identification of molecular bonding.

2.5. Electrochemical Characterization

The electrochemical properties of NiMoO$_4$ NPs and NiMoO$_4$/3D-rGO electrodes were evaluated with three-electrode cells in a 3 M-KOH electrolyte with the potentiostat system (VersaStat 4 with VersaStudio, Sundsvall, Sweden). A reference electrode (saturated Hg/Hg$_2$Cl$_2$) and platinum (counter electrode) were used for supercapacitor measurement.

The electrochemical properties of the NiMoO$_4$ NPs and NiMoO$_4$/3D-rGO hybrid nanocomposite were characterized using cyclic voltammetry (CV), galvanostatic charge–discharge (GCD), and continuous charge–discharge tests. According to the battery-like behaviour of the Ni-based materials, calculation of the capacity (Cg^{-1}) is the true assessment of energy stored on such electrodes, rather than the capacitance with the unit of Fg^{-1} [7]. The specific capacity of the electrodes can be obtained from the discharge time of both cyclic voltammetry curves and the galvanostatic charge/discharge curves, according to the following equations [7]:

$$Q_{CV} = \frac{1}{mv} \int I dV, \tag{1}$$

$$Q_{GCD} = \frac{\int I dt}{m}, \tag{2}$$

where Q (Cg^{-1}) is the charge stored [7], I (A) refers to the discharge current, Δt (s) is attributed to the discharge time, and m (g) is related to the mass loading of the active material on the current collector [26,27].

3. Results and Discussion

3.1. Investigation of Morphology and Structural Characterization

FTIR spectral analysis was performed to investigate the chemical bonds of the produced samples and reduction of GO during the process. Figure 1 illustrates the FTIR spectra of the pure GO, NiMoO$_4$ NPs, and NiMoO$_4$/3D-rGO nanocomposite. The FTIR spectrum of GO showed a characteristic peak of C=O corresponding to stretching vibrations from a carbonyl group at ~1765 cm^{-1}. The bond at ~1650 cm^{-1} was attributed to the stretching of aromatic C=C bonds and the peaks at ~1397 cm^{-1}, ~1208 cm^{-1}, and ~1038 cm^{-1} corresponded to the C–OH and the deformation of C–O–C and C–O bonds, respectively [28,29]. As can be seen, the FTIR spectrum containing a broad bond between ~2800 and ~3600 cm^{-1} was due to the hydroxyl (O–H) stretching vibration mode of water molecules adsorbed on the surface of the samples [30]. The bonds before 1000 cm^{-1} were related to metal oxides. The characteristic bonds of NiMoO$_4$ appearing at 444 cm^{-1} and 750 cm^{-1} were assigned to Mo–O–Mo and Mo–O–Ni vibrations, respectively. The bond appearing at 836 cm^{-1} can be attributed to the symmetric stretching of Mo=O bond [24]. As can be seen in the FTIR spectrum of NiMoO$_4$/3D-rGO nanocomposite, most of the oxygenated groups, like C–OH, C–O, and C=O, disappeared from the FTIR spectrum of the NiMoO$_4$/3D-rGO nanocomposites and C–O–C peak was weakened. This means that GO sheets were reduced to rGO, during the attraction of Ni^{2+} to the oxygen groups of GO with a negative charge [31]. In addition, the reduction of the strong absorption peak at 3150 cm^{-1} demonstrated the successful chemical reaction between graphene oxide and NiMoO$_4$ to form the NiMoO$_4$/3D-rGO nanocomposite.

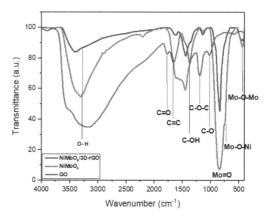

Figure 1. Fourier transform infrared spectrometry (FTIR) spectra of the graphene oxide (GO) sheets, NiMoO$_4$ nanoparticles (NPs), and NiMoO$_4$/3D-rGO nanocomposite.

X-ray powder diffraction analysis was performed to investigate the structure of the synthesized powders. In Figure 2, the XRD patterns of the NiMoO$_4$ NPs, NiMoO$_4$/3D-rGO nanocomposite, and pristine GO is illustrated. Diffraction peaks of the NiMoO$_4$ NPs in Figure 2a match with the code number of 00-045-0142. It should be noted that the peaks at around 27.3°, 30.36°, and 45° were attributed to α-NiMoO$_4$ [32]. During the process, Ni^{2+} and MoO$_4{}^{2-}$ cations reacted with OH$^-$ to form a Ni–Mo hydroxide precursor, as shown in the following equations [33]:

$$NO_3{}^- + 7H_2O + 8e^- \rightarrow NH_4{}^+ + 10OH^-, \tag{3}$$

$$Ni^{2+} + Mo^{6+} + 8OH^- \rightarrow NiMo(OH)_8, \tag{4}$$

$$NiMo(OH)_8 \rightarrow NiMoO_4 + 4H_2O. \tag{5}$$

Although the results indicate the presence of a dominant NiMoO$_4$ phase, small quantities of the orthorhombic Mo$_{17}$O$_{47}$ phase with a space group of Pba2 (JCPDS No. 01-071-0566) were also observed in the structure. The peaks at 2θ = 21.12° (510), 22.67° (001), and 32.07° (521) could be related to the Mo$_{17}$O$_{47}$ phase. It can be concluded that Mo$_{17}$O$_{47}$ phase consisted of 17 MoO$_3$ phases, which some of the oxygens replaced with vacancies, and finally Mo$_{17}$O$_{47}$ was formed instead of Mo$_{17}$O$_{51}$. In Figure 2b, the XRD pattern of the NiMoO$_4$/3D-rGO nanocomposite can be seen. No obvious peak related to GO (at 2θ = 10.6° (001) in Figure 2c) could be found in the XRD pattern of the NiMoO$_4$/3D-rGO nanocomposite. This confirms that the oxygen-containing functional groups of graphene oxide could be removed by reduction through NiMoO$_4$ NPs at 350 °C; furthermore, the rGO peaks did not appear in the XRD pattern of NiMoO$_4$/3D-rGO nanocomposite, which may be due to the lesser content of rGO and the strong NiMoO$_4$ NP peaks covering the rGO peaks [34,35]. Moreover, the distance between the graphene sheets was fully filled with the insertion of NiMoO$_4$ NPs into the interplanar groups [36].

Another reason for this is that carbon materials like rGO are amorphous and cannot be detected by X-ray diffraction.

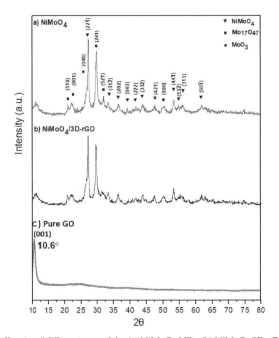

Figure 2. X-ray diffraction (XRD) patterns of the (**a**) NiMoO$_4$ NPs, (**b**) NiMoO$_4$/3D-rGO nanocomposite after heating at 350 °C, and (**c**) pristine GO.

In addition, the XRD pattern of the NiMoO$_4$ powder revealed an asymmetric peak at $2\theta = 27.3°$. For a more accurate study, this peak was opened. As shown in Figure 3, after deconvolution of the asymmetric peak, it was divided into two distinct peaks, at $2\theta = 26.86°$ and $2\theta = 27.31°$. These peaks were respectively attributed to MoO$_3$ (040) and α-NiMoO$_4$ (221$^-$) phases. The MoO$_3$ phase was produced from a combination of Mo^{6+} ions and oxygen.

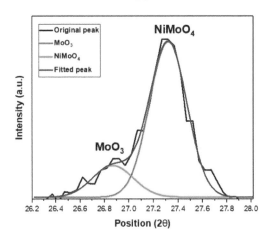

Figure 3. Deconvolution of asymmetric peak of the X-ray diffraction pattern of the NiMoO$_4$ nanopowder at $2\theta = 27.3°$.

The Raman spectra of GO, NiMoO$_4$ NPs, and NiMoO$_4$/3D-rGO nanocomposites (Figure 4) were analyzed to check whether GO was reduced during the synthesis. It was established from Figure 4a that GO had two characteristic peaks: the D bond at 1347 cm^{-1} originated from the defects in the hexagonal graphitic network and the G bond at 1584 cm^{-1} was due to the vibration of the sp^2 domain of carbon atoms [22]. Comparison of the Raman spectra of NiMoO$_4$/3D-rGO nanocomposite (Figure 4b) with GO spectra exhibited a shift in the D (1383.87 cm^{-1}) and G (1580.56 cm^{-1}) positions. The intensity ratio of the D and G peaks (I$_D$/I$_G$) reveals the degree of the disorders. By reduction of graphene oxide and removal of oxygen groups, the defects increased and the I$_D$/I$_G$ ratio showed an increase from 0.98 to 1.02. This was attributed to the increase of the defects in rGO [22,25,37]. The intensive peak at 969 cm^{-1}, and some slight intensity peaks at 928, 825, 394, and 325 cm^{-1} denoted the stoichiometric α-phase NiMoO$_4$ (Figure 4b,c) [32,38,39]. The peaks at 969 and 928 cm^{-1} were related to the symmetric and asymmetric stretching modes of the Mo=O bond, whereas the peak at 825 cm^{-1} occurred due to the Ni–O–Mo symmetric stretch. The peaks at 394 cm^{-1} and 325 cm^{-1} were representative of the bending band of Mo–O and Mo–O–O [32,37,40]. The Raman results confirmed the reduction of GO to rGO in the NiMoO$_4$/3D-rGO nanocomposite. Thus, by increasing the defects in rGO, more active sites can be provided for enhancing the electron storage and improvement of capacity [41].

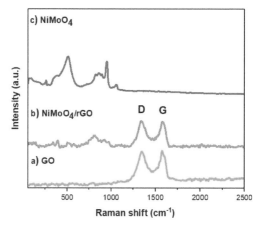

Figure 4. Raman spectrum of the GO, NiMoO$_4$, and NiMoO$_4$/3D-rGO nanocomposites.

Scanning electron microscopy (SEM) was performed to investigate the morphology and the porosities of the nanocomposite. The SEM images and EDX characterisation of the NiMoO$_4$ NPs and NiMoO$_4$/3D-rGO nanocomposite are illustrated in Figure 5. According to Figure 5a,b, NiMoO$_4$ NPs possess nanorod morphology with three-dimensional orientation and lots of porosities among the nanorods. To some extent, negligible agglomeration can be seen, which is related to the magnetic nature of NiMoO$_4$ [14]. Figure 5c,d demonstrates the decoration and dispersibility of the NiMoO$_4$ nanorods on the surface of rGO sheets. The attachment of metal ions to the functional group of GO sheets through electrostatic interactions resulted in the growth of NiMoO$_4$ NPs on the 3D reduced GO sheets [29,34]. Therefore, it can be stated that the rGO sheets play important roles as a useful site for the growth of NiMoO$_4$ NPs. The formation of porosities in the NiMoO$_4$/3D-rGO structure confirmed the transformation of starch to CO$_2$ and CO gases during the thermal process. The elemental compositions of the NiMoO$_4$ and NiMoO$_4$/3D-rGO samples were also analyzed by the energy dispersive X-ray method. Figure 5e,f shows the EDX spectra of NiMoO$_4$/3D-rGO and NiMoO$_4$ samples. The peaks confirm the presence of Ni, Mo, O, and C elements in the NiMoO$_4$/3D-rGO nanocomposite and Ni, Mo, and O elements in the NiMoO$_4$ spectrum.

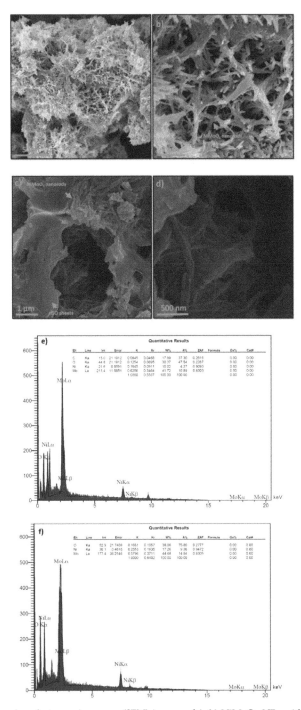

Figure 5. Scanning electron microscopy (SEM) images of (**a,b**) NiMoO₄ NPs with two different magnifications, (**c,d**) NiMoO₄/3D-rGO nanocomposite with various magnifications, (**e**) corresponding EDX spectra of NiMoO₄/3D-rGO, and (**f**) corresponding EDX spectra of NiMoO₄.

3.2. Electrochemical Measurement of the NiMoO$_4$ NPs and NiMoO$_4$/3D-rGO Electrodes

The potentiostat analysis system was used to explore the electrochemical performance of NiMoO$_4$ NPs and NiMoO$_4$/3D-rGO electrodes. Figure 6a and b show the CV curves of pure NiMoO$_4$ and NiMoO$_4$/3D-rGO electrodes. The working cell potential range of the electrodes was in the potential window of 0–0.7 V during the CV measurements at different scan rates from 10 to 100 mVs^{-1}. Figure 6a exhibits the CV patterns of NiMoO$_4$ NPs electrode. According to the curves, as the scan rate increased, the anodic and cathodic peaks moved towards the higher and lower potentials, respectively. This is related to the electrode internal resistance, which limits the charge transfer at high scan rates. At lower scan rates, a pair of redox peaks could be seen in each CV curve, indicating the Faradic reaction of Ni(II)/Ni(III) [26,42] due to the reversible phase change between Ni(OH)$_2$ and NiOOH (0.44 V and 0.18 V) [43]. By scanning the electrode, the OH$^-$ ions will be dispersed on the surface of Ni(OH)$_2$ layer and form the NiOOH phase, as shown in Equation (6):

$$Ni(OH)_2 + OH^- \leftrightarrow NiOOH + H_2O + e^-. \tag{6}$$

Mo is not involved in redox reactions. This element, with a high conductivity, can increase the whole capacity of the cell [26,44]. As shown in Figure 6b, the CV curves of the NiMoO$_4$/3D-rGO electrode, similar to the NiMoO$_4$ electrode (Figure 6a), at low scan rates displayed two redox peaks related to NiMoO$_4$ faradaic reactions (Ni^{2+}/Ni^{3+}). The NiMoO$_4$/3D-rGO hybrid electrode consisted of two mechanisms for storing energy: the faradaic reaction of NiMoO$_4$ as well as the formation of an electric double-layer because of the presence of the graphene oxide sheets in the composite. This can lead to the storage of more electrons in the NiMoO$_4$/3D-rGO in comparison to NiMoO$_4$. According to Equation (1), the maximum capacity for a NiMoO$_4$/3D-rGO electrode is 249.8 Cg^{-1} (356.4 Fg^{-1}) at a scan rate of 10 mVs^{-1} and at the scan rates of 20, 50, and 100 mVs^{-1}, it was calculated to be equal to 212.8 Cg^{-1} (304 Fg^{-1}), 152.95 Cg^{-1} (218.5 Fg^{-1}), and 84 Cg^{-1} (120 Fg^{-1}), respectively, whereas the calculated capacity for NiMoO$_4$ at the scan rates of 10, 20, 50, and 100 mVs^{-1} is 211.2 Cg^{-1} (301.7 Fg^{-1}), 192.85 Cg^{-1} (275.5 Fg^{-1}), 126.84 Cg^{-1} (181.2 Fg^{-1}), and 67.69 Cg^{-1} (96.7 Fg^{-1}). This shows that the introduction of rGO to NiMoO$_4$ increased the capacity. Figure 6c illustrates the CV curves of Ni foam, NiMoO$_4$ NPs, and the NiMoO$_4$/3D-rGO nanocomposite. As can be seen, the CV curve of the NiMoO$_4$/3D-rGO nanocomposite had a slightly larger enclosed area at the scan rate of 10 mV·s^{-1}, reflecting that the NiMoO$_4$/3D-rGO hybrid nanocomposite electrode has higher specific capacity and the Ni foam capacity can be neglected. More capacity of NiMoO$_4$/3D-rGO confirms that oxygenated functional groups of GO were removed and more conductivity provided, resulting in more capacity. In addition, the differences between the reduction and oxidation potentials for the NiMoO$_4$/3D-rGO and NiMoO$_4$ electrodes were 241 mV and 280 mV, respectively. Thnis reveals an improvement in the electrochemical reversibility of the NiMoO$_4$/3D-rGO electrode due to the addition of rGO sheets to NiMoO$_4$ NPs.

The galvanostatic charge–discharge (GCD) measurements were carried out at different current densities (1–50 A g^{-1}) and the potential range of 0–0.5 V for NiMoO$_4$ NPs and NiMoO$_4$/3D-rGO electrodes (Figure 6d,e). A pair of voltage plateaus as well as nonlinear curves of the GCD graphs confirm the faradaic redox reaction of Ni^{2+}/Ni^{3+} ions of the electrodes [45]. The NiMoO$_4$/3D-rGO electrode exhibited a specific capacity of 100 Cg^{-1} (200 Fg^{-1}) at the high current density of 10 A g^{-1}, whereas the specific capacity of pure NiMoO$_4$ NPs was 80.2 Cg^{-1} (160.4 Fg^{-1}) at 10 A g^{-1}.

Stability over repeated charge and discharge cycles is a critical parameter for supercapacitors. The cyclability of NiMoO$_4$ and NiMoO$_4$/3D-rGO electrodes was tested by continuous charge–discharge measurements (Figure 6f) over 400 cycles (24 h) at the current density of 2 A g^{-1} in 3M-KOH aqueous solution as an electrolyte. As shown in Figure 6f, increasing the number of cycles from 1 to 400 resulted in a reduction in the specific capacity of the NiMoO$_4$ electrode from 314.8 Cg^{-1} (629.6 Fg^{-1}) to 179.4 Cg^{-1} (358.8 Fg^{-1}). The specific capacity decreased during the first several cycles, then it remained stable at approximately 179.4 Cg^{-1} (358.8 Fg^{-1}), losing less than 57% capacity by the end of the test. The capacity of the NiMoO$_4$/3D-rGO electrode decreased with the increase in the cycle number, which can be related to the decrease in the electroactive sites. Then, the specific capacity remained stable for the next cycles. In the end, approximately 41% of the initial capacity was still maintained. Electrochemical measurements showed that specific capacity of both electrodes reached around 160 Cg^{-1} (320 Fg^{-1}) at the current density of 2 A g^{-1} after 24 h. It can be concluded that the nanorod morphology of the $NiMoO_4$ and the porous structure facilitate better penetration and migration through the electrode surface path.

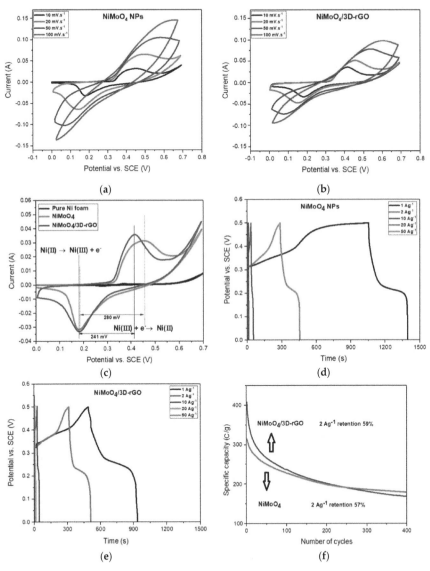

Figure 6. Cyclic voltametry curves of (**a**) pure $NiMoO_4$ NP electrode and (**b**) hybrid $NiMoO_4$/3D-rGO electrode at various scan rates (10–100 mV·s^{-1}); (**c**) CV curves of Ni foam, $NiMoO_4$ NPs, and $NiMoO_4$/3D-rGO electrode (10 mV·s^{-1}); Galvanostatic charge discharge curves of (**d**) pure $NiMoO_4$ NPs electrode and (**e**) hybrid $NiMoO_4$/3D-rGO electrode at different current densities (1–50 A g^{-1}); (**f**) age test of hybrid $NiMoO_4$/3D-rGO and pure $NiMoO_4$ NPs electrode at a current density of 2 A g^{-1}.

The capacitance of the NiMoO$_4$/3D-rGO electrode was estimated to be around 900 Fg^{-1} at 1 A g^{-1}, which is a relatively good result, compared to many other studies; for example, the C/NiMoO$_4$ electrode (859 Fg^{-1} at 1 A g^{-1}) [15], the NiMoO$_4$/CoMoO$_4$ nanorods electrode (1445 Fg^{-1} at 1 A g^{-1}) [16], the CoMoO$_4$–NiMoO$_4$·xH$_2$O electrode (1039 Fg^{-1} at 2.5 mA cm^{-2}) [21], the NiMoO$_4$ electrode (800 Fg^{-1} at 1 A g^{-1}) [22], and the sphere-shaped NiMoO$_4$ electrode (1000 Fg^{-1} at 1 A g^{-1}) [25].

4. Conclusions

A NiMoO$_4$/3D-rGO nanocomposite was prepared with a simple and effective precipitation method in starch media. This technique is useful for synthesizing a large amount of NiMoO$_4$/3D-rGO nanocomposite for supercapacitor application. The results confirmed the transformation of GO to rGO during the process. The calculated specific capacity of the NiMoO$_4$/3D-rGO and NiMoO$_4$ electrode was 450 Cg^{-1} (900 Fg^{-1}) and 350 Cg^{-1} (700 Fg^{-1}), respectively, at the current density of 1 A g^{-1}. The good specific capacity of the nanocomposite can be attributed to the 3D structure of the rGO sheets and the faradaic reaction of NiMoO$_4$ nanoparticles. This porous structure resulted in rapid electron and ion transportation. Therefore, using starch to produce a porous surface area is recommended to synthesize the nanocomposite of NiMoO$_4$ and rGO for the investigation of hybrid supercapacitor electrodes.

Author Contributions: S.A.R.: Investigation, synthesis, and writing—original draft, R.S.M.: Supervision, N.B.: Editing and formal analysis, M.P.: Software, and H.O.: Supervision and editing manuscript. All authors have read and agreed to the published version of the manuscript.

Funding: This research received no external funding.

Acknowledgments: The authors acknowledge Tarbiat Modares University and Mid Sweden University for providing the facilities and technical assistance for this research. We thank the honorable supervisors, Magnus Hummelgård, and all of the personnel who work in the research.

Conflicts of Interest: The authors declare no conflict of interest.

References

1. Xia, X.; Zhang, Y.; Chao, D.; Guan, C.; Zhang, Y.; Li, L.; Fan, H.J. Solution synthesis of metal oxides for electrochemical energy storage applications. *Nanoscale* **2014**, *6*, 5008–5048. [CrossRef] [PubMed]
2. Iro, Z.S.; Subramani, C.; Dash, S.S. A brief review on electrode materials for supercapacitor. *Int. J. Electrochem. Sci.* **2016**, *11*, 10628–10643. [CrossRef]
3. Budhiraju, V.S.; Kumar, R.; Sharma, A.; Sivakumar, S. Structurally stable hollow mesoporous graphitized carbon nanofibers embedded with NiMoO$_4$ nanoparticles for high performance asymmetric supercapacitors. *Electrochim. Acta* **2017**, *238*, 337–348. [CrossRef]
4. Gund, G.S.; Lokhande, C.D.; Park, H.S. Controlled synthesis of hierarchical nanoflake structure of NiO thin film for supercapacitor application. *Alloys Compd.* **2018**, *741*, 549–556. [CrossRef]
5. Wang, H.; Song, Y.; Liu, W.; Yan, L. Three dimensional Ni(OH)$_2$/rGO hydrogel as binder-free electrode for asymmetric supercapacitor. *Alloys Compd.* **2018**, *735*, 2428–2435. [CrossRef]
6. Liu, F.; Chu, X.; Zhang, H.; Zhang, B.; Su, H.; Jin, L.; Yang, W. Synthesis of self-assembly 3D porous Ni(OH)$_2$ with high capacitance for hybrid supercapacitors. *Electrochim. Acta* **2018**, *269*, 102–110. [CrossRef]
7. Brisse, A.L.; Stevens, P.; Toussaint, G.; Crosnier, O.; Brousse, T. Ni(OH)$_2$ and NiO Based Composites: Battery Type Electrode Materials for Hybrid Supercapacitor Devices. *Materials* **2018**, *11*, 1178. [CrossRef]
8. Chiu, H.Y.; Cho, C.P. Mixed-Phase MnO$_2$/N-Containing Graphene Composites Applied as Electrode Active Materials for Flexible Asymmetric Solid-State Supercapacitors. *Nanomaterials* **2018**, *8*, 924. [CrossRef]
9. Fei, H.; Saha, N.; Kazantseva, N.; Moucka, R.; Cheng, Q.; Saha, P. A highly flexible supercapacitor based on MnO$_2$/RGO nanosheets and bacterial cellulose-filled gel electrolyte. *Materials* **2017**, *10*, 1251. [CrossRef]
10. Prakash, N.G.; Dhananjaya, M.; Narayana, A.L.; Shaik, D.P.; Rosaiah, P.; Hussain, O.M. High Performance One Dimensional α-MoO$_3$ Nanorods for Supercapacitor Applications. *Ceram. Int.* **2018**, *44*, 9967–9975. [CrossRef]
11. Zhang, H.; Zhou, Y.; Ma, Y.; Yao, J.; Li, X.; Sun, Y.; Li, D. RF magnetron sputtering synthesis of three-dimensional graphene@Co$_3$O$_4$ nanowire array grown on Ni foam for application in supercapacitors. *Alloys Compd.* **2018**, *740*, 174–179. [CrossRef]

12. Zhao, J.; Li, Z.; Yuan, X.; Yang, Z.; Zhang, M.; Meng, A.; Li, Q. A High-Energy Density Asymmetric Supercapacitor Based on Fe_2O_3 Nanoneedle Arrays and $NiCo_2O_4/Ni(OH)_2$ Hybrid Nanosheet Arrays Grown on SiC Nanowire Networks as Free-Standing Advanced Electrodes. *Adv. Energy Mater.* **2018**, *8*, 1702787–1702791. [CrossRef]

13. Zhang, C.; Lei, C.; Cen, C.; Tang, S.; Deng, M.; Li, Y.; Du, Y. Interface polarization matters: Enhancing supercapacitor performance of spinel $NiCo_2O_4$ nanowires by reduced graphene oxide coating. *Electrochim. Acta* **2018**, *260*, 814–822. [CrossRef]

14. Umapathy, V.; Neeraja, P.; Manikandan, A.; Ramu, P. Synthesis of $NiMoO_4$ nanoparticles by sol–gel method and their structural, morphological, optical, magnetic and photocatlytic properties. *Trans. Nonferrous Met. Soc. China* **2017**, *27*, 1785–1793. [CrossRef]

15. Xuan, H.; Xu, Y.; Zhang, Y.; Li, H.; Han, P.; Du, Y. One-step combustion synthesis of porous $CNTs/C/NiMoO_4$ composites for high-performance asymmetric supercapacitors. *Alloys Compd.* **2018**, *745*, 135–146. [CrossRef]

16. Nti, F.; Anang, D.A.; Han, J.I. Facilely synthesized $NiMoO_4/CoMoO_4$ nanorods as electrode material for high performance supercapacitor. *Alloys Compd.* **2018**, *742*, 342–350. [CrossRef]

17. Yesuraj, J.; Elumalai, V.; Bhagavathiachari, M.; Samuel, A.S.; Elaiyappillai, E.; Johnson, P.M. A facile sonochemical assisted synthesis of α-$MnMoO_4$/PANI nanocomposite electrode for supercapacitor applications. *Electroanal. Chem.* **2017**, *797*, 78–88. [CrossRef]

18. Xiao, K.; Xia, L.; Liu, G.; Wang, S.; Ding, L.X.; Wang, H. Honeycomb-like $NiMoO_4$ ultrathin nanosheet arrays for high-performance electrochemical energy storage. *J. Mater. Chem. A* **2015**, *3*, 6128–6135. [CrossRef]

19. Xiong, X.; Ding, D.; Chen, D.; Waller, G.; Bu, Y.; Wang, Z.; Liu, M. Three-dimensional ultrathin $Ni(OH)_2$ nanosheets grown on nickel foam for high-performance supercapacitors. *Nano. Energy* **2015**, *11*, 154–161. [CrossRef]

20. Mohapatra, D.; Parida, S.; Singh, B.K.; Sutar, D.S. Importance of microstructure and interface in designing metal oxide nanocomposites for supercapacitor electrodes. *Electroanal. Chem.* **2017**, *803*, 30–39. [CrossRef]

21. Liu, M.C.; Kong, L.B.; Lu, C.; Ma, X.J.; Li, X.M.; Luo, Y.C.; Kang, L. Design and synthesis of $CoMoO_4$–$NiMoO_4 \cdot xH_2O$ bundles with improved electrochemical properties for supercapacitors. *J. Mater. Chem. A* **2013**, *1*, 1380–1387. [CrossRef]

22. Liu, T.; Chai, H.; Jia, D.; Su, Y.; Wang, T.; Zhou, W. Rapid microwave-assisted synthesis of mesoporous $NiMoO_4$ nanorod/reduced graphene oxide composites for high-performance supercapacitors. *Electrochim. Acta* **2015**, *180*, 998–1006. [CrossRef]

23. Kianpour, G.; Salavati-Niasari, M.; Emadi, H. Sonochemical synthesis and characterization of $NiMoO_4$ nanorods. *Ultrason. Sonochem.* **2013**, *20*, 418–424. [CrossRef] [PubMed]

24. De Moura, A.P.; de Oliveira, L.H.; Rosa, I.L.; Xavier, C.S.; Lisboa-Filho, P.N.; Li, M.S.; Varela, J.A. Structural, optical, and magnetic properties of $NiMoO_4$ nanorods prepared by microwave sintering. *Sci. World J.* **2015**, *1*, 1–9. [CrossRef]

25. Jinlong, L.; Miura, H.; Meng, Y. A novel mesoporous $NiMoO_4$@rGO nanostructure for supercapacitor applications. *Mater. Lett.* **2017**, *194*, 94–97. [CrossRef]

26. Arshadi Rastabi, S.; Sarraf Mamoory, R.; Dabir, F.; Blomquist, N.; Phadatare, M.; Olin, H. Synthesis of $NiMoO_4$/3D-rGO Nanocomposite in Alkaline Environments for Supercapacitor Electrodes. *Crystals* **2019**, *9*, 31. [CrossRef]

27. Kumar, Y.; Kim, H.J. Effect of Time on a Hierarchical Corn Skeleton-Like Composite of CoO@ZnO as Capacitive Electrode Material for High Specific Performance Supercapacitors. *Energies* **2018**, *11*, 3285. [CrossRef]

28. Ossonon, B.D.; Bélanger, D. Synthesis and characterization of sulfophenyl-functionalized reduced graphene oxide sheets. *RSC Adv.* **2017**, *7*, 27224–27234. [CrossRef]

29. Srivastava, M.; Uddin, M.E.; Singh, J.; Kim, N.H.; Lee, J.H. Preparation and characterization of self-assembled layer by layer $NiCo_2O_4$–reduced graphene oxide nanocomposite with improved electrocatalytic properties. *Alloys Compd.* **2014**, *590*, 266–276. [CrossRef]

30. Fayer, M.D. *Ultrafast Infrared Vibrational Spectroscopy*; CRC Press: New York, NY, USA, 2013; Volume 4, p. 128.

31. Li, Y.; Jian, J.; Fan, Y.; Wang, H.; Yu, L.; Cheng, G.; Sun, M. Facile one-pot synthesis of a $NiMoO_4$/reduced graphene oxide composite as a pseudocapacitor with superior performance. *RSC Adv.* **2016**, *6*, 69627–69633. [CrossRef]

32. Bankar, P.K.; Ratha, S.; More, M.A.; Late, D.J.; Rout, C.S. Enhanced field emission performance of NiMoO$_4$ nanosheets by tuning the phase. *Appl. Surf. Sci.* **2017**, *418*, 270–274. [CrossRef]

33. Ezeigwe, E.R.; Khiew, P.S.; Siong, C.W.; Tan, M.T. Synthesis of NiMoO$_4$ nanorods on graphene and superior electrochemical performance of the resulting ternary based composites. *Ceram. Int.* **2017**, *43*, 13772–13780. [CrossRef]

34. Azarang, M.; Shuhaimi, A.; Yousefi, R.; Sookhakian, M. Effects of graphene oxide concentration on optical properties of ZnO/RGO nanocomposites and their application to photocurrent generation. *Appl. Phys.* **2014**, *116*, 84307–84313. [CrossRef]

35. Jamali-Sheini, F.; Azarang, M. Effect of annealing temperature and graphene concentrations on photovoltaic and NIR-detector applications of PbS/rGO nanocomposites. *Ceram. Int.* **2016**, *42*, 15209–15216. [CrossRef]

36. Uddin, A.S.M.I.; Phan, D.T.; Chung, G.S. Synthesis of ZnO nanoparticles-reduced graphene oxide composites and their intrinsic gas sensing properties. *Surf. Rev. Lett.* **2014**, *21*, 1450086–1450097. [CrossRef]

37. Guan, X.H.; Lan, X.; Lv, X.; Yang, L.; Wang, G.S. Synthesis of NiMoSO/rGO Composites Based on NiMoO$_4$ and Reduced Graphene with High-Performance Electrochemical Electrodes. *ChemistrySelect* **2018**, *3*, 6719–6728. [CrossRef]

38. Jothi, P.R.; Kannan, S.; Velayutham, G. Enhanced methanol electro-oxidation over in-situ carbon and graphene supported one dimensional NiMoO$_4$ nanorods. *J. Power Sources* **2015**, *277*, 350–359. [CrossRef]

39. Ghosh, D.; Giri, S.; Das, C.K. Synthesis, characterization and electrochemical performance of graphene decorated with 1D NiMoO$_4$·nH$_2$O nanorods. *Nanoscale* **2013**, *5*, 10428–10437. [CrossRef]

40. Jothi, P.R.; Shanthi, K.; Salunkhe, R.R.; Pramanik, M.; Malgras, V.; Alshehri, S.M.; Yamauchi, Y. Synthesis and Characterization of α-NiMoO$_4$ Nanorods for Supercapacitor Application. *Eur. J. Inorg. Chem.* **2015**, *22*, 3694–3699. [CrossRef]

41. Liu, H.; Zhang, G.; Zhou, Y.; Gao, M.; Yang, F. One-step potentiodynamic synthesis of poly (1,5-diaminoanthraquinone)/reduced graphene oxide nanohybrid with improved electrocatalytic activity. *J. Mater. Chem. A* **2013**, *1*, 13902–13913. [CrossRef]

42. Li, Y.; Jian, J.; Xiao, L.; Wang, H.; Yu, L.; Cheng, G.; Sun, M. Synthesis of NiMoO$_4$ nanosheets on graphene sheets as advanced supercapacitor electrode materials. *Mater. Lett.* **2016**, *184*, 21–24. [CrossRef]

43. Trafela, Š.; Zavašnik, J.; Šturm, S.; Rožman, K.Ž. Formation of a Ni(OH)$_2$/NiOOH active redox couple on nickel nanowires for formaldehyde detection in alkaline media. *Electrochim. Acta* **2019**, *309*, 346–353. [CrossRef]

44. Jiang, G.; Zhang, M.; Li, X.; Gao, H. NiMoO$_4$@Ni(OH)$_2$ core/shell nanorods supported on Ni foam for high-performance supercapacitors. *RSC Adv.* **2015**, *85*, 69365–69370. [CrossRef]

45. Yedluri, A.K.; Anitha, T.; Kim, H.J. Fabrication of Hierarchical NiMoO$_4$/NiMoO$_4$ Nanoflowers on Highly Conductive Flexible Nickel Foam Substrate as a Capacitive Electrode Material for Supercapacitors with Enhanced Electrochemical Performance. *Energies* **2019**, *12*, 1143. [CrossRef]

Article

Energy Management of a DC Microgrid Composed of Photovoltaic/Fuel Cell/Battery/Supercapacitor Systems

Ahmed A. Kamel [1], Hegazy Rezk [2,3,*], Nabila Shehata [1] and Jean Thomas [4]

[1] Faculty of postgraduate studies for Advanced Sciences, Beni-Suef University, Beni Suef 62511, Egypt; ahmed.abdeltawab2014@gmail.com (A.A.K.); nabila.shehata@yahoo.com (N.S.)

[2] College of Engineering at Wadi Addawaser, Prince Sattam bin Abdulaziz University, Alkharj 11942, Saudi Arabia

[3] Electrical Engineering Department, Faculty of Engineering, Minia University, Minia 61519, Egypt

[4] Electrical Department, Faculty of Engineering, Beni-Suef University, Beni Suef 62511, Egypt; jhh_thomas@yahoo.com

* Correspondence: hr.hussien@psau.edu.sa; Tel.: +966-547-416-732

Received: 26 June 2019; Accepted: 10 September 2019; Published: 19 September 2019

Abstract: In this paper, a classic proportional–integral (PI) control strategy as an energy management strategy (EMS) and a microgrid stand-alone power system configuration are proposed to work independently out of grid. The proposed system combines photovoltaics (PVs), fuel cells (FCs), batteries, and supercapacitors (SCs). The system supplies a dump load with its demand power. The system includes DC/DC and DC/AC converters, as well as a maximum power point tracking (MPPT) to maximize the harvested energy from PV array. The system advantages are represented to overcome the problem of each source when used individually and to optimize the hydrogen consumption. The classic PI control strategy is used to control the main system parameters like FC current and the state-of-charge (SOC) for the battery and SC. In order to analyze and monitor the system, it was implemented in the MATLAB/Simulink. The simulation done for fuzzy logic and high frequency decoupling and state machine control strategies to validate the PI classic control strategy. The obtained results confirmed that the system works efficiently as a microgrid system. The results show that the SOC for the battery is kept between 56 and 65.4%, which is considered a proper value for such types of batteries. The DC bus voltage (VDC) is kept within the acceptable level. Moreover, the H_2 fuel consumption is 12.1 gm, as the FCs are used as supported sources working with the PV. A big area for improvement is available for cost saving, which suggests the need for further research through system optimization and employing different control strategies.

Keywords: DC microgrid; energy management; hybrid power system; energy efficiency

1. Introduction

In power system grids, the microgrid is identified as a distributed energy system (DES), including generators, energy storage elements like batteries (B) and supercapacitors to balance the generated power and the consumed power [1–3], an energy management system to control the entire operation of the microgrid sources [4,5], and load. All of these items are combined together and work in parallel with the utility grid, or out of grid as a stand-alone system used for a small area and few consumers [6,7]. Generally, the microgrid is considered a cluster of the utility grid [8], as shown in Figure 1. Using a utility grid for power distribution has some disadvantages, such as transmission losses, especially when the generating plants are far away from the consumers, bad environmental impact because of emission, and climate change due to the use of conventional resources in the generation phase. Microgrids represent an alternative option that has the potential to overcome these problems. Microgrids can minimize power losses through transmission, reduce CO_2 emissions, and limit climate change, especially when using renewable energy resources in electricity generation. It also saves money in several ways, such as preventing outage, selling electricity to national grids through feed, and tariff or net metering systems. Using power systems that combine renewable sources with zero emission besides energy storage elements makes the system able to achieve maximum efficiency compared to systems containing conventional sources [9–11]. An energy management system is a system that controls the operation of the microgrid (MG) system. It uses many approaches and control strategies to get maximum system performance. These control strategies may include a state machine, a classic proportional–integral (PI) control, a fuzzy logic control, an external energy maximization, an equivalent minimization, and a frequency decoupling control strategy [12].

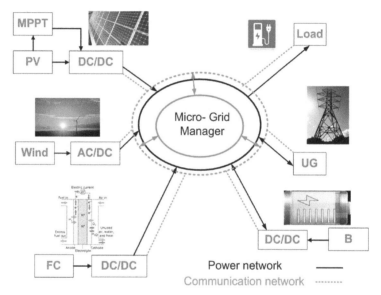

Figure 1. Configuration of DC microgrid.

Manoj et al. [13] discussed the two main types of MG: Alternative current microgrid (ACMG) and direct current microgrid (DCMG). The DCMG has some advantages over the ACMG, such as high efficiency, easy connection on the DC bus, and system reliabilies. They confirmed that in DCMGs, there are three factors influencing power disturbance: Fluctuation of power exchange, power variation between the storage system and the power sources, and the fast change in the DC bus load. Ravichandrudu et al. [14] designed an MG system using renewable sources, which has the benefits of utilizing renewable energy sources and reducing transmission losses when using wind and hydro. The proposed microgrid system has three operation modes. Phurailatpam et al. [15] proposed a DCMG system that includes a photovoltaic (PV) power system and uses the battery as an energy storage system. The DC/DC converters were discussed, as well as the maximum power point tracking (MPPT) for the PV. The system performance at constant and variable values for irradiance, wind speed, and load was monitored and analyzed. The simulation results showed that the system maintained the DC bus voltage at constant value, which confirmed the advantages of the DCMG compared with the ACMG. Elsied et al. [16] proposed a novel energy management strategy based on binary particle swarm optimization (BPSO) to optimize the performance of the MG, maximize the micro grid power, and decrease the system CO_2 emissions. The system was supported by an experimental lab test. The obtained results proved that the BPSO is efficient when used with the MG. For a DCMG hybrid system, Garita et al. [17] examined the efficiency of an energy management strategy (EMS) used in a DCMG configuration containing a PV and a battery integrated together in one unit. An energy management system was used with three main system configurations. It works through seven operation modes as power flow direction under two different case studies (OFF—grid and peak shaving) to achieve maximum system performance. In their study of DCMG, Eghtedarpour et al. [18] propose three levels of control to improve the performance of the DCMG. The first level is where no communication is required and the control is done based on the local measurements. The second level is based on a DC microgrid energy management system. The third level is the top level of control, which controls multi-microgrids. Shehata et al. [19] proposed an energy management strategy for DCMG based on a multi-agent system applied using the JADE framework, where PI controllers are used as an EMS. The interface between the multi-agent system and the MATLAB/Simulink software was done through the MACSimJX interface. An analytical solution used as a reference model supported by a numerical method was proposed by Hadj-Said et al. [20] to confirm the suitability of the proposed EMS used in parallel hybrid electric vehicles (HEV) to achieve maximum system performance. The proposed EMS was applied successfully on continuous and discrete optimization cases. Sedaghati et al. [21] discussed a PV–FC–B–SC hybrid system based on grid-connected microgrids. A control strategy called adaptive fractional fuzzy sliding mode control (AFFSMC) was used for the inverter. The fuzzy rules are designed to accurately estimate the uncertain parameters. The results showed that the proposed strategy works efficiently. When the analytical solution was applied for a certain model, such as the optimal auxiliary functions method (OAFM) proposed by Herisanu et al. [22], it was found to be a reliable and efficient tool for mechanical and electrical performance of the system. For the PV/wind hybrid system in a microgrid, the wind turbine generator is characterized by its slow response, while the PV array enjoys a fast response. If both generators are combined in one system, the voltage of the DC bus takes a long time to reach a stable condition that affects the overall system response, especially when the load is variably switching [23]. For the battery/supercapacitor hybrid system, as proposed by Vasily et al. [24], a number of the internal problems of the battery negatively affects the system's overall performance, such as its short life cycle and its low efficiency due to the number of charging/discharging times, causing fast breakdown of the battery. For the PV/FC hybrid power system, the main problem is the efficiency of both the PV and the FC, and the low density and high initial cost of the PV arrays. Despite the concerns raised for the mentioned hybrid systems, sometimes the configuration itself cannot achieve its target, as [25], when all the system components are connected in a series and, as a result, if there is a problem in any component (PV, Electrolyzer, storage, and FC), the power production could be affected, thus impacting the system's financial return and cost. Using the PV as a preliminary source decreases the unwanted side effects of using the FC, such as low efficiency during the operation period, high cost, unstable low generated voltage, and finally, the high ripple current linked to the output voltage which reduces the FC lifetime [26,27].

Along the same lines, a number of studies explored the same application with the same configuration; however, compared with the present paper, it was found that all the previous works that have the same components connected the SC to the DC bus through a DC/DC converter to control the charging/discharging of the SC [28–30]. By contrast, in the present paper, the SC has been directly connected to the DC, which has two advantages over other connection methods. The first one is reducing the cost of the system through removing one DC/DC converter, and the second advantage is the fast response of the SC to load changing. One of the important parameters for the proton exchange membrane fuel cell (PEMFC) used in this paper is that it works at low temperatures, which gives the chance to enhance its size to achieve maximum power production. The solid oxide fuel cell (SOFC) used for the hybrid system [21], by contrast, needs long start-up times and requires insulation and heat dispersion due to temperature concerns. In addition, the PEMFC has higher efficiency, more fuel flexibility, smaller size as it does not require cooling or thermal dissipation, and is less expensive compared to SOFC. The main contribution of this paper is to propose an MG configuration containing PV/FC/B/SC to supply a dump AC variable load with its needed power. It also proposes a classic PI control strategy as an EMS to control the FC current and calculate the hydrogen consumed by the fuel cell. For the proposed system, the PV system decreases the FC hydrogen consumption during day light when it is available, especially in areas of high irradiance values, for a long time. It also gives a chance to increase the size of the clean power generated to decrease greenhouse gasses and global warming, which is good for the environment and climate change. Also, using a battery and a supercapacitor as storage elements gives the system an advantage over individual systems, as the specific power is low, while the specific energy is high—but the supercapacitor has a high specific power and low specific energy [31,32]. In this paper, the performance of the system is simulated using two different energy management strategies (fuzzy logic control strategy and high frequency decoupling and state machine control strategies) to validate the proposed PI control strategy. The simulation results show that the PI control strategy is better than the high frequency decoupling and state machine control strategy in terms of hydrogen consumption. Although the hydrogen consumed by the fuzzy logic control strategy is close to the consumed value by PI, the PI control strategy is easy and simple for implementation. The overall system performance proves that the system works efficiently when applied for a three-phase AC variable load. The cost optimization issue was not calculated in this study, and could be considered in future papers. The next sections of this paper present the overall system description of the proposed system structure, system components, the control strategy, results and discussion, and finally, the conclusion.

2. Overall System Description

A hybrid power system containing a PV, FC, battery, and SC is designed to examine the optimal configuration for the power system shown in Figure 2. It is implemented and simulated in MATLAB/Simulink (version 2018a) software to monitor, control, and analyze the system. The PV panel is connected to a DC/DC boost converter, which is controlled through an MPPT based on perturb and observe (P&O) to get the maximum power of the solar panels. The irradiance values were assumed through the Simulink signal builders and the temperature is fixed at 25 °C. The system component was chosen to get the maximum system performance when working together to cover the disadvantages of each source when operating as a stand-alone source. The system is designed to supply a dump load with its demand power. The system covers the disadvantages of each individual power system when working as a separate system. The surplus power from PV is utilized in charging the SC and battery. It keeps their state-of-charge (SOC) in proper value for operation.

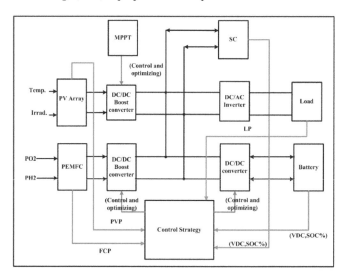

Figure 2. Block diagram of the proposed system.

3. System Components

3.1. PV Arrays

The equivalent circuit for the PV solar cell is represented in Figure 3 [33,34], where I_L is the current generated inside the solar cell according to the sunlight. As a basic configuration, the solar cell is a P–N junction, so the diode current I_D is taken into consideration.

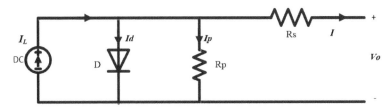

Figure 3. Equivalent circuit for solar cell.

The output current is calculated by applying Kirchhoff's law and is displayed as follows:

$$I = I_L - I_0 \left(\exp\left(\frac{(Vo + IR_s)}{aV_T} \right) - 1 \right) - \frac{Vo + IR_s}{R_{SH}} \tag{1}$$

where I denotes the output current; I_0 is the diode saturation current; V denotes the output voltage; a is the identifying factor of the cells; and V_T is called the thermal voltage [35,36]. Solving Equation (1) for short circuit current and R_P value is very large compared to the series resistance. The saturation current as a function of temperature is calculated directly by the following equation [37]:

$$I_0 = I_{0.ref} \left(\frac{T_{ref}}{T} \right)^3 \exp\left\{ \frac{qE_g}{ak} \left(\frac{1}{T_{ref}} - \frac{1}{T} \right) \right\} \tag{2}$$

where T_{ref} and T are the reference and ambient temperatures, respectively, and E_g denotes the energy gap of the material. The relation between the photon current generated and the solar irradiance is represented as follows [38]:

$$I_L = \frac{G}{G_a} \left(I_{L.ref} + V_{sc}\, \Delta T \right) \tag{3}$$

where: G is the solar irradiance; G_a is the reference solar irradiance, which is equivalent to 1 KW/m^2 at the standard test condition (STC); ΔT denotes the temperature difference between the actual temperature and the temperature at STC, which is 25 °C; and V_{sc} is the temperature coefficient [39]. Rezk et al. represent in [40] the modeling of the *I–V* curve of the PV under all conditions of irradiance and temperature. Table 1 represents the data sheet of the solar panels used in the proposed system in this paper.

Table 1. Data sheet for the TPB 156x156-72-P-295W.

Module Type	TPB 156x156-72-P-295W
Module power class	295 Wp
Composition	72 (156 × 156 mm) polycrystalline silicon solar cells per module
Max. power (Pmpp) (in W)	295
Max. voltage (Umpp) (in V)	35.3
Max. current (Impp) (in A)	8.36
Open-circuit voltage (Uoc) (in V)	44.3
Short-circuit current (ISC) (in A)	8.67
Cell temperature (TNOCT) (in °C)	46
Module efficiency (in %)	15.2

Figure 4 shows the relationship between the generated current and output power, along with the voltage for the type TPB 156x156-72-P-295W PV solar module used in this paper. The maximum value of the current, called I_{SC} current, when the output terminals are shorted and the maximum value of voltage is the V_{oc} at the value of zero current.

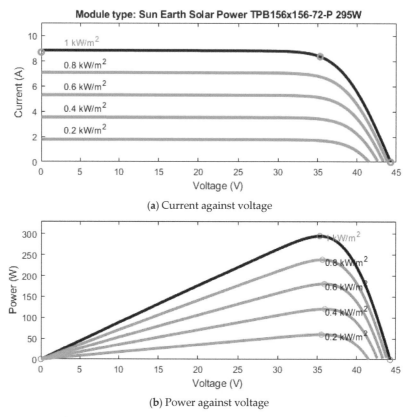

(a) Current against voltage

(b) Power against voltage

Figure 4. Characteristics of TPB 156x156-72-P-295W PV solar module.

Several researches discussed in detail the MPPT algorithms and how to use them for maximizing the generated power from PV [41–45]. In this paper, the MPPT used is based on the perturb and observe (P&O) algorithm with the flowchart shown in Figure 5. P&O is the most common technique in which the power is compared at several samples and perturbs the current. This process is repeated until the difference in power is zero. Larminie et al. discussed several techniques of MPPT, including P&O [46].

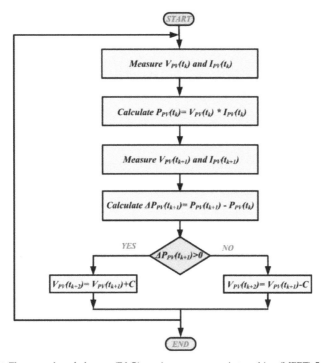

Figure 5. The perturb and observe (P&O) maximum power point tracking (MPPT) flowchart.

3.2. Fuel Cell

The advantages of the proton exchange membrane fuel cell (PEMFC) include its high efficiency, reaching up to 45%, high energy density out of small dimensions (up to 2 W/cm²), silent operation, low-temperature operation, fast start-up, and system robustness [47,48]. The most important advantage of the PEMFC is the minimum pollutants, where the hydrogen fuel used in FC has no adverse effects on the environment [49,50]. In spite of the FC advantages, the FC has some disadvantages like its slow response to the load variation, its unstable output voltage, its short lifetime because of the increase in current ripple, and its relatively high cost. The overall equivalent circuit for the FC discussed by Outeiro et al. is shown in Figure 6 [50].

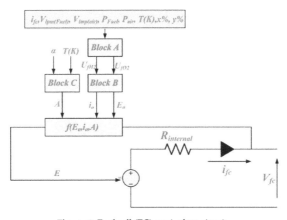

Figure 6. Fuel cell (FC) equivalent circuit.

The modeling and simulation of the FC were discussed in [51] as follows:

$$A = \frac{RT}{z\alpha F} \tag{4}$$

$$E_{oc} = E_n K_c \tag{5}$$

$$i_o = \frac{zFk\left(P_{H_2} + P_{O_2}\right)}{RH} e^{\frac{-\Delta G}{RT}} \tag{6}$$

where R is 8.3145 J/(mol K); K_c denotes voltage constant at nominal condition of operation; T is temperature of operation (K); F is 96,485 A s/mol; ΔG is size of the activation barrier, which depends on the type of electrode and catalyst used; h denotes Planck's constant = 6.626×10^{-34} J s; z is number of moving electrons; k denotes Boltzmann's constant = 1.38×10^{-23} J/K; P_{O_2} denotes partial pressure of oxygen inside the stack (atm); P_{H_2} is partial pressure of hydrogen inside the stack (atm); E_n is Nernst voltage, which is the thermodynamics voltage of the cells and depends on the temperatures and partial pressures of reactants and products inside the stack (V); and α is charge transfer coefficient, which depends on the type of electrodes and catalysts used. For block A in the FC equivalent circuit, the utilization factor for the fuel and the air (H_2 and O_2) is calculated as follow:

$$U_{fO_2} = \frac{n^r_{O_2}}{n^{in}_{O_2}} = \frac{6000RTN_{ifc}}{2zFP_{air}V_{lmp(air)} \, y\%} \tag{7}$$

$$U_{fH_2} = \frac{n^r_{H_2}}{n^{in}_{H_2}} = \frac{6000RTN_{ifc}}{zFP_{fuel}V_{lmp(fuel)} \, x\%} \tag{8}$$

where P_{fuel} is absolute supply pressure of fuel (atm); P_{air} is absolute supply pressure of air (atm); N denotes number of cells; $V_{lpm(air)}$ is air flow rate (L/min); $V_{lpm(fuel)}$ is fuel flow rate (L/min); y denotes percentage of oxygen in the oxidant (%); x denotes percentage of hydrogen in the fuel (%); and the 60,000 constant comes from the conversion from the L/min flow rate used in the model to m^3/s (1 L/min = 1/60,000 m^3/s). The Nernst voltage is determined in Block B as follows:

$$E_{Nernst \, t} = 1.229 - (T - 298.15)\frac{-44.43}{zF} + \frac{-RT}{zF}\ln\left(P_{H_2}P^{\frac{1}{2}}_{O_2}\right), \text{ when } T \leq 100 \,^\circ\text{C} \tag{9}$$

$$E_{Nernst \, t} = 1.229 - (T - 298.15)\frac{-44.43}{zF} + \frac{-RT}{zF}\ln\left(\frac{P_{H_2}P^{\frac{1}{2}}_{O_2}}{P_{H_2O}}\right), \text{ when } T > 100 \,^\circ\text{C} \tag{10}$$

The partial pressure for H_2, O_2, and H_2O are calculated also in block B as follows:

$$P_{H_2} = \left(1 - U_{fH_2}\right)x\% \, P_{fuel} \tag{11}$$

$$P_{H_2O} = \left(W + 2y\%U_{fO_2}\right)P_{air} \tag{12}$$

$$P_{O_2} = \left(1 - U_{fO_2}\right)y\% \, P_{air} \tag{13}$$

where W denotes percentage of water vapor in the oxidant (%) and P_{H_2O} is partial pressure of water vapor inside the stack (atm). The updated values of the exchange current i_o and the open circuit voltage are calculated according to the partial pressure and the Nernst voltage. The (I–V) curve of the FC used in this paper is represented in Figure 7. It represents the relation between the current density and the FC voltage. Table 2 represents the data sheet for FC used in present case study.

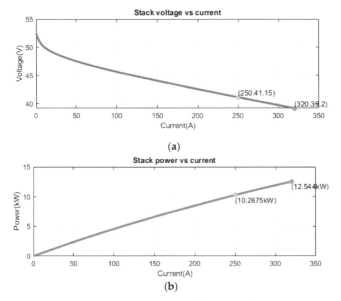

Figure 7. *I–V* curve for the FC used in proposed system.

Table 2. Data sheet parameters of the proton exchange membrane fuel cell (PEMFC).

Parameters	
Nominal power (w)	10,287.5
Max power (w)	12,544
Nernst voltage (V)	1.1491
Hydrogen (H_2)	98.98%
Oxidation (O_2)	42.885
Fuel flow rate (lpm)	114.9
Air flow rate (lpm)	732
System Temp (K)	318
P Fuel (bar)	1.16
P Air (bar)	1

3.3. Battery

The main target of using and integrating a battery and a supercapacitor with the renewable energy resources power configuration is to store the surplus of energy produced from the other sources and reuse it whenever there is a shortage in energy [52,53]. In this paper, the proposed design was built based on the lithium-ion battery. The battery parameters are shown in Table 3 and the discharge parameters are shown in Table 4.

Table 3. Battery parameters.

Parameters	
Nominal voltage (V)	48
Rated capacity (Ah)	40
Initial state-of-charge (%)	65

Table 4. Battery discharge parameters.

Discharge	
Maximum capacity (Ah)	40
Fully charged voltage (V)	55.8714
Nominal discharge current (A)	17.3913
Capacity (Ah) at nominal voltage	36.1739
Exponential zone [Voltage (V), Capacity (Ah)]	[52.3, 1.96]
Discharge current [i_1, i_2, i_3, …] (A)	[20, 80]

For batteries, there are three main modeling types: Mathematical model, electro-chemical model, and equivalent circuit model [54]. Honorat et al. discussed the three methods of fast characteristics of automotive lithium-ion second life batteries [55], whereas madani et al. [56] discussed the electrical equivalent circuit for second order batteries. Jiuchun et al. [57] argue that the electrical equivalent circuit is the best model for representing the battery, due to the unsuitability of the mathematical model for actual application and the complexity of the electro-chemical model. Valant et al. [58] tested the modules used in secondary application of grid in lab conditions. Generally, the equivalent circuit of the ideal battery combines the open circuit voltage and the battery internal resistance. Figure 8 represents the equivalent battery circuit and Figure 9 represents the battery performance—where V_b is the battery voltage; i_b is the battery current; Voc is the open circuit voltage as a function of SOC h (t); R_p and C_p are the resistance and the capacitance of the battery polarization; and R_s is the internal resistance.

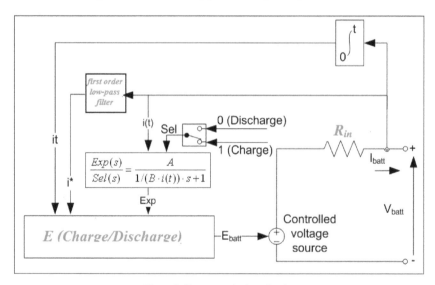

Figure 8. Battery equivalent circuit.

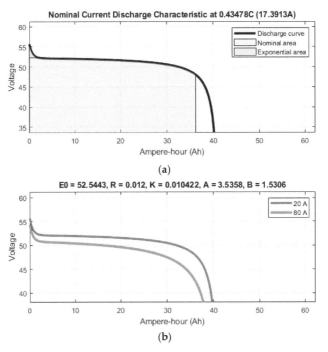

Figure 9. The battery performance.

For discharge mode, the battery voltage equation is represented as follow [59]:

$$V_{batt} = E_0 - K\left(\frac{Q}{Q-it}\right) i^* - K\left(\frac{Q}{Q-it}\right) it + A e^{-Bit} \ (i^* > 0) \tag{14}$$

For charge mode, the battery voltage equation is represented as follows:

$$V_{batt} = E_0 - K\left(\frac{Q}{it+0.1\,Q}\right) i^* - K\left(\frac{Q}{Q-it}\right) it + A e^{-Bit} \ (i^* < 0) \tag{15}$$

The fully charged state voltage is displayed as follows:

$$V_{full} = E_0 - Ri + A \tag{16}$$

The exponential section voltage is calculated as follows:

$$V_{exp} = E_0 - K\left(\frac{Q}{Q-Q_{exp}}\right)(Q_{exp}+i) - Ri + A e^{\frac{-3}{Q_{exp}} Q_{exp}} \tag{17}$$

And finally, the nominal zone cell voltage is calculated as follows:

$$V_{nom} = E_0 - K\left(\frac{Q}{Q-Q_{nom}}\right)(Q_{nom}+i) - Ri + A e^{\frac{-3}{Q_{exp}} Q_{nom}} \tag{18}$$

where E_0 is constant voltage, in V; K is polarization constant, in Ah^{-1}; i^* is low frequency current dynamics, in A; i is battery current, in A; it is extracted capacity, in Ah; Q is maximum battery capacity, in Ah; A is exponential voltage, in V; and B is exponential capacity, in Ah^{-1}. SOC is estimated according to coulomb counting by accumulating the capacity during battery charging/discharging:

$$SOC = SOC_0 - \frac{1}{Q_n} \int_0^t \eta i dt \tag{19}$$

3.4. Supercapacitors (SC)

The SC is used with the battery to decrease the peak current in the battery when the load is highly fluctuating because of its high specific power. As the battery cannot supply the needed power at a high rate because of it characteristics, the SC covers this power shortage. SC has a high-efficiency cycle (about 100%) which is suitable for both frequent charge/discharge cycles and storage of energy, compared to the battery, which is used to supply the average needed power. It means that the SC delivers the power faster and has more charge/recharge cycles than the battery [60]. That is why the supercapacitor is used as a complementary element with other electrical sources that have different dynamic behavior and different energy storage quantities [61]. The main parameters of the supercapacitor source used in this paper are shown in Table 5, while the self-discharge parameters are listed in Table 6.

Table 5. Supercapacitor (SC) main parameters.

Parameters	
Rated capacitance (F)	15.6
Equivalent DC series resistance (Ohms)	1.50×10^{-1}
Rated voltage (V)	291.6
Initial voltage (V)	270
Operating temperature (Celsius)	25

Table 6. Supercapacitor (SC) self-discharge parameters.

Self-Discharge	
Current prior open-circuit (A)	10
Voltage at 0 s, 10 s, 100 s, and 1000 s $[V_{oc}, V_3, V_4, V_5]$ (V)	[48, 47.8, 47.06, 44.65]
Charge current $[i_1, i_2, i_3, \ldots]$ (A)	[10, 20, 100, 500]

Figure 10 shows the electrical equivalent circuit for the SC [62], where C is the capacitance of SC, EPR is the equivalent parallel resistances, and ESR is the equivalent series internal resistances of SC.

$$E_{UC} = \frac{1}{2} C \left(V_i^2 - V_f^2 \right) E_{UC} \tag{20}$$

where E_{UC} is the dragged energy from the SC and $\left(V_i^2 - V_f^2 \right)$ is the voltage change between the final and initial voltage. The series/parallel configuration of the capacitors in SC was determined by the value of the terminal voltage. The total capacitance and resistance for the SC can be determined as follows:

$$R_{total} = n_s \frac{ESR}{n_p} \tag{21}$$

$$ESR = \frac{\Delta V_d}{I_d} \tag{22}$$

$$C_{total} = n_p \frac{C}{n_s} \tag{23}$$

$$C = I_d \frac{(t_2 - t_1)}{(V_2 - V_1)} \tag{24}$$

where I_d is the discharging current; n_s is the number of connected capacitors in series; and n_p is the number of series strings in parallel. The characteristics of the supercapacitor charge is shown in Figure 11.

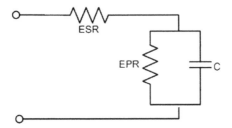

Figure 10. Electrical equivalent circuits for the SC.

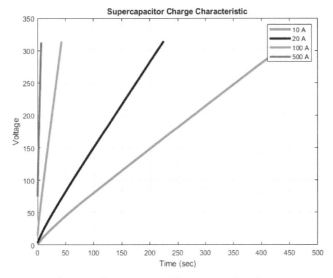

Figure 11. Characteristics of the supercapacitor charge.

4. Energy Management System

As the proposed system in this paper contains multi-electrical power sources like FC and PV, an energy storage element could be used as a power source when discharging. Hence, there is a need for an energy management strategy, based on a computer program to control, monitor, and optimize the system operation to get maximum system performance [63]. The EMS is used to increase the system overall efficiency, decrease the hydrogen fuel consumption in the FC power system when the FC is used as a component of the hybrid system, increase the life cycle of the system component to keep the stability of the DC voltage, and control the SOC and prevent its deep discharge [64]. There are many EMSs and control strategies that are used with renewable hybrid power systems, such as state machine control, fuzzy logic control (FLC), Control loop cascade, proportional–integral–derivative (PID) control approach, and instantaneous optimization approach. The PI cascaded control was promoted in this paper as a microgrid control strategy. This control strategy calculates and sets the reference values of FC current, battery charge, and discharge currents. Figure 12 shows the flowchart of the fuel cell current control with each comparison step between the reference values of SOC, load power with PV power, and finally, the minimum and maximum values for the DC bus voltage (VDC). Regarding the

EMS used in the present case study, the classic PI controller gives the simplest way because it has a few number of inputs, is easy to be configured, has feedback, and is inexpensive compared with other EMSs like FLC. FLC has more rules for more accuracy, low speed performance, and much more cost with regard to the programming and hardware interface [65].

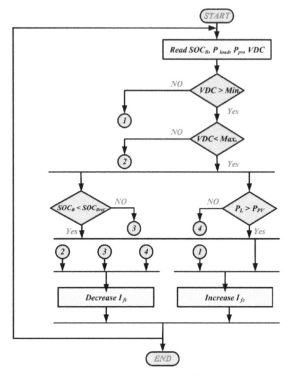

Figure 12. Fuel cell current control flow chart.

The PI control strategy is simulated in MATLAB/Simulink, as illustrated in Figure 13. Figure 13a shows the actual percentage of SOC measured in the simulation program compared with a predetermined value (60%). According to the difference, the program increases/decreases the reference power of the fuel cell. Figuré 13b illustrates the shortage power from load/PV comparison, where the behavior of the strategy output is calculated according to the difference between the load and the PV power. Figure 13c shows the reference value of the fuel cell current according to the reference power, and Figure 13d shows the fuel cell reference current according to VDC limits.

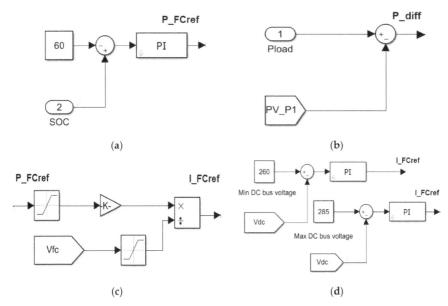

(a)   (b)

(c)   (d)

Figure 13. (**a**) State-of-charge (SOC) % check, (**b**) power shortage, (**c**) fuel cell current, (**d**) fuel cell reference current according to DC bus voltage (VDC).

The control of the charging/discharging battery current is done according to PI control strategy. It depends on a summation of two PI controllers. One of them is based on the power difference between the load and the PV power, and the other is based on the difference between the actual and the reference of the VDC, which is 270 VDC in this case. According to the result of these two PI controllers, the output of the strategy by charge/discharge the battery. Figure 14 shows the flowchart of the battery charge/discharge procedure.

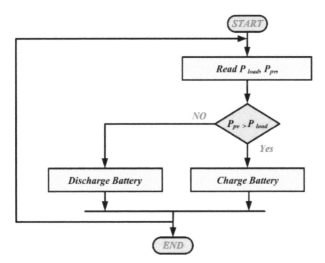

Figure 14. Battery charge/discharge flowchart according P_{load} and P_{pv}.

Figure 15 shows the PI control strategy MATLAB model for battery.

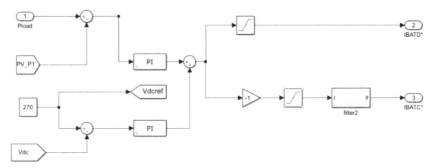

Figure 15. PI control strategy for battery charge/discharge.

The system is designed to supply a variable three-phase dump load, as shown in Figure 16a, with its needed power. The load profile is assumed to be variable with different levels of power in order to test the performance of the proposed system at variable value of load (from about 0 to 9 kW) along the total duration of simulation. The system is implemented and simulated in MATLAB/Simulink for a total simulation time of 350 s. During this period, the behavior of the system could be divided into three different stages according to the PV power, as shown in Table 7.

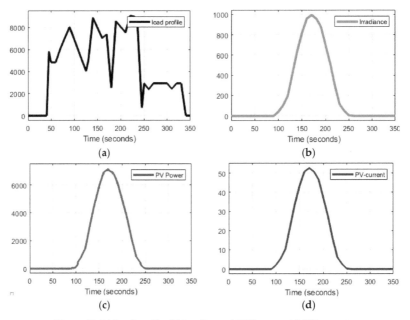

Figure 16. (a) Load profile, (b) irradiance, (c) PV power, (d) PV current.

Table 7. Simulation stages and its boundaries.

Stages	Time	PV Power (Watt)
Stage #1	From 0 to 90 s	0
Stage #2	From 90 to 250 s	0 > PV > 7326
Stage #3	From 250 to 350 s	0

4.1. Performance during Stage #1

The start of this stage is at 0 s, and its end is determined at 90 s. In this period, the PV power is zero, as shown in Figure 16c. In this stage, the zero PV power is because of no solar irradiance, as shown in Figure 16b. During the period from 0 to 40 s, there is a surplus of power because the load is zero. The fuel cell current is at minimum (i_{fc} = 20 A) because the SOC initial value (60%) is greater than its reference value (60%), which leads the VDC to increase more than 270 V, which is the set point for the system DC bus voltage. The battery and supercapacitor start to charge, as per battery charging current (Figure 17a).

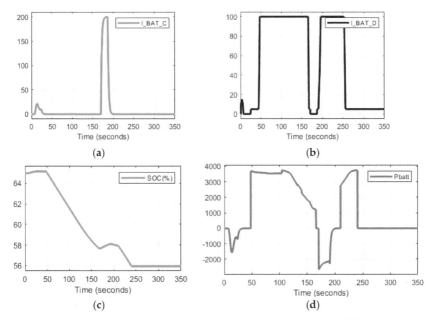

Figure 17. (a) Battery charging current, (b) battery discharging current, (c) battery SOC%, (d) battery power.

Figure 18 shows line voltage, phase current, fuel cell power, and battery power. At 40 s, the load starts to increase and the supercapacitor starts to supply the load with the power faster than the fuel cell and battery because of its charge/discharge response. Then, the battery and fuel cell start to supply their power at 43 s and 44 s, as illustrated in Figure 18c,d. The fuel cell power, the battery, and the supercapacitor power follows the load power until the end of this stage at 90 s, according to the PI control strategy.

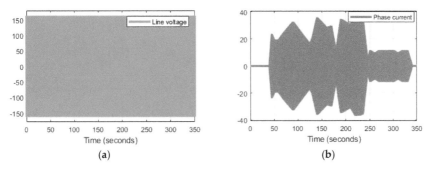

Figure 18. *Cont.*

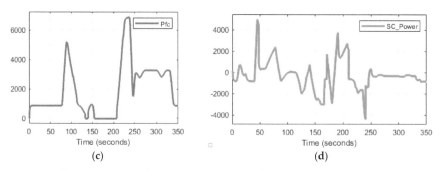

Figure 18. (a) Line voltage, (b) phase current, (c) fuel cell power, (d) SC Power.

4.2. Performance during Stage #2

This stage is initiated at 90 s and ends at 250 s. The PV output power increases with the increasing of the irradiance value according to day light. As the temperature is assumed to be constant ($t = 25\,°C$), the PV power is mainly dependent on the PV current shown in Figure 16d. The EMS estimates the difference between the load power P_{load} and the PV power, and then determines the new values of the fuel cell current and the battery charge/discharge current. In this case, the PV power interferes in controlling the battery charging/discharging, as well as controlling the FC current (Figure 19a).

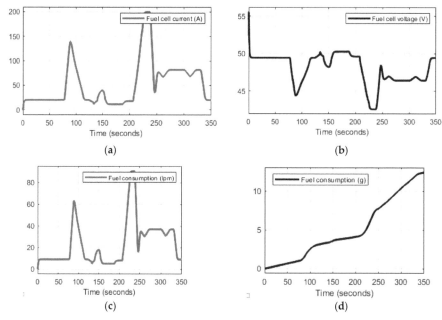

Figure 19. (a) Fuel cell current, (b) fuel cell voltage, (c) fuel cell rate, (d) fuel consumption.

The PV power reaches its maximum value of 7326 watt at 170 s. At this point, the generated PV power exceeds the load demand, leading to a surplus of power. This surplus power is used to charge the battery and supercapacitor through the PI control strategy, and simultaneously decreases the power consumed from the fuel cell at the same time until 180 s. During the period from 80 to 250 s, the load demand is more than the generated PV power and there is no surplus power. The battery and supercapacitor then discharge and share the load power with the PV array and fuel cell. Figure 20 shows the results for all of the system collected together.

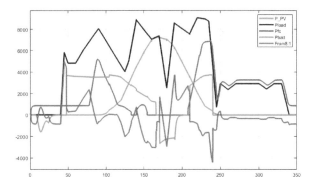

Figure 20. All power sources along the simulation period.

4.3. Performance during Stage #3

This stage starts at 250 s up until 350 s. In this stage, the generated PV power is zero and the load is supplied by the fuel cell power, battery power, and the supercapacitor power, the same as stage 1. At the end of this stage, the load power is zero and the fuel cell supplies power for charging the battery and supercapacitor. As a result, the VDC increases again and reaches the value of 285 VDC by the end of this stage. The system keeps working in this manner until the end of the simulation. Figure 20 shows the performance of all power sources along the overall simulation period from 0 to 350 s.

5. Comparison Study

In order to validate the results of the classic PI control strategy as a satisfying energy management system to control hybrid energy sources working in a microgrid, the results of a classic PI control strategy simulation were compared with the results of a fuzzy control strategy and a high frequency decoupling and state machine control strategy. Qi Li et al. [66] discussed the fuzzy logic control strategy as an energy management system for a FC/battery/supercapacitor hybrid vehicle. It was used for enhancing the fuel to increase the mileage of the journey, and the results show that the system achieved the power requirement at four standard driving cycles. In the current research, the FIS function in MATLAB was configured with three input signals to the FIS, which were load power, PV power, and battery SOC%, while the output was configured as the FC current. Eight rules were assigned to represent the operation of the system during the simulation periods. Figure 21 shows the Fuzzy FIS configuration used in the simulation.

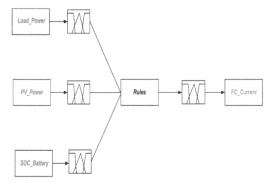

Figure 21. The fuzzy inference system (FIS) function configuration.

Table 8 shows the eight rules of the fuzzy control system which control the fuel cell current output signal. The eight rules contain the most conditions so that the system can act accordingly.

Table 8. Fuzzy control system rules.

Rule	Load	PV	SOC	IFC
1	Low	Low	OK	Low
2	OK	Low	OK	OK
3	OK	OK	OK	Low
4	High	Low	OK	High
5	High	OK	OK	Low
6	High	High	OK	Low
7	High	High	Low	Low
8	Low	Low	High	Low

High frequency decoupling control strategy is used to decrease the effect of the transient load change by insulating the PEMFC current from the high frequency transient load change [67]. The configuration of the frequency decoupling and state machine control strategy implemented in this paper is shown in Figure 22.

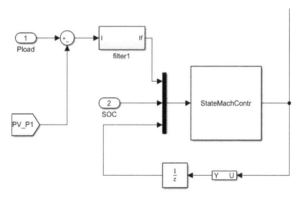

Figure 22. Frequency decoupling and state machine configuration.

Figure 23 displays the results of the fuzzy logic control strategy, where Figure 23a shows the hydrogen consumption along the simulation period. Figure 23b shows the SOC of the battery.

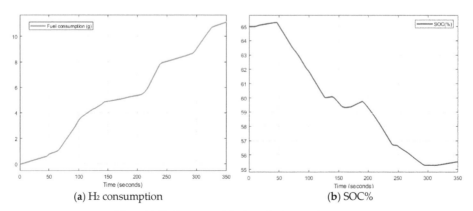

(a) H₂ consumption

(b) SOC%

Figure 23. Performance of the fuzzy logic control strategy.

Figure 24 displays the results of the high frequency decoupling and state machine control strategy. The hydrogen consumption along the simulation period is 19.9 (g), as shown in Figure 24a. Figure 24b shows the SOC of the battery.

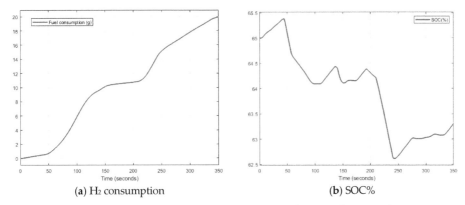

(a) H₂ consumption   (b) SOC%

Figure 24. Performance of the high frequency decoupling and state machine control strategy.

For the main two parameters, hydrogen consumption and SOC (%), Figure 25 shows the comparison between three control strategies. Figure 25a shows the fuel consumption for PI, fuzzy, and high frequency decoupling and state machine. The high frequency decoupling and state machine is the biggest strategy for hydrogen consumption, while the fuzzy and the PI are close.

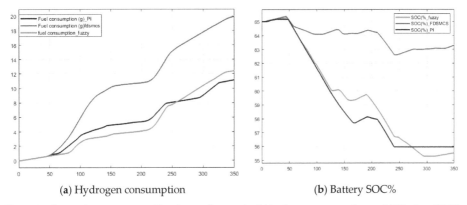

(a) Hydrogen consumption   (b) Battery SOC%

Figure 25. Comparison among considered control strategies (a) hydrogen consumption and (b) battery SOC%.

Table 9 summarizes the comparison among the classic PI control strategy, fuzzy logic control strategy, and high frequency decoupling and state machine control strategy.

Table 9. Comparison among considered control strategies.

Method	PI	Fuzzy	High Frequency Decoupling and State Machine
H₂ consumption (gram)	12.13	13	19.9
SOC%	56–65	55–65	63.19–65.39

6. Conclusions

The microgrid combined with a renewable hybrid power system is a very promising, efficient, and clean power generation system. It can replace the conventional fuels easily. In this study, the obtained results prove that the designed hybrid power system—which combines PV, FC, battery, and SC—works efficiently at decreasing the effects of the FC disadvantages. It solves the problems of the individual source and supplies the load with sufficient and stable power. The PV array supplies the main power and the FC compensates for the power shortage because of shading and night time. The battery and SC are used to solve the problems of slow response of the FC during the fast change of the load power and to remove the peak power from the system. In cases where surplus power exists, this power is used to charge the battery to keep battery SOC% at a healthy level (between 57 and 65.4%), and sustain the VDC within the range of 265 to 285 in good condition. Moreover, the H_2 fuel consumption is 12.13 gm, as the FC is used as supported sources working with the PV. The system was simulated for another two control strategies, fuzzy and high frequency decoupling state machines. The results for comparison prove that the PI control strategy is better than the high frequency decoupling state machine. In addition, for the fuzzy control strategy, although the results were close, the PI is easier for implementation. Future researches could focus on improving the system with regard to cost optimization. It is also suggested that an electric electrolyzer should be attached to the system to use the surplus power in hydrogen production. Further researches could focus on helping environment interests such as global warming and climate change, in addition to using other EMS strategies and optimization techniques to improve the system overhaul performance.

Author Contributions: Conceptualization, H.R. and A.A.K.; methodology, H.R. and A.A.K.; software, H.R. and A.A.K.; validation, H.R. and A.A.K.; formal analysis, H.R. and A.A.K.; investigation, H.R. and A.A.K.; resources, H.R., A.A.K., N.S. and J.T.; writing—original draft preparation, H.R. and A.A.K.; writing—review and editing, H.R., A.A.K., N.S. and J.T.

Funding: This research received no external funding.

Conflicts of Interest: The authors declare no conflict of interest.

Nomenclature

MG	Microgrid
ACMG	Alternative current microgrid
DCMG	Direct current microgrid
PEMFC	Proton exchange membrane fuel cell
PV	Photovoltaic
HESS	Hybrid energy storage system
B	Battery
BESS	Battery energy storage system
I_{bat}	Battery current (A)
MPPT	Maximum power point tracking
OAFM	Optimal auxiliary function method
O&P	Observe and perturb
VDC	Voltage in direct current side (V)
SCADA	Supervisory control and data acquisition
EMS	Energy management system
P_{bat}	Battery power (KW)
JADE	Java agent development framework
BPSO	Binary particle swarm optimization
AFFSMC	Adaptive fractional fuzzy sliding mode control
CO_2	Carbon dioxide (gm)
PH	Positive high
PL	Positive low
P_{load}	Load demand (KW)
Psc	Supercapacitor power (KW)
P_{sur}	Surplus power (KW)
i_{FC}	Full cell current (A)
REHS	Renewable energy hybrid system

SC	Supercapacitor
ICA	Imperialist competitive algorithm
PSO	Particle swarm optimization
QPSO	Quantum behaved particle swarm optimization
ACO	Ant colony optimization
COA	Cuckoo optimization algorithm
PCM	Power control management
SOC	State-of-charge (%)
VDC	DC bus voltage (V)
SOC_{bat}	State-of-charge of battery (%)
SOCsc	State-of-charge of supercapacitor (%)

References

1. El-Shahat, A.; Sumaiya, S. DC-Microgrid System Design, Control, and Analysis. *Electronics* **2019**, *8*, 124. [CrossRef]
2. Einan, M.; Hossein, T.; Mahdi, P. Optimized fuzzy-cuckoo controller for active power control of battery energy storage system, photovoltaic, fuel cell and wind turbine in an isolated micro-grid. *Batteries* **2017**, *3*, 23. [CrossRef]
3. García-Quismondo, E.; Almonacid, I.; Martínez, M.A.C.; Miroslavov, V.; Serrano, E.; Palma, J.; Salmerón, J.P.A. Operational Experience of 5 kW/5 kWh All-Vanadium Flow Batteries in Photovoltaic Grid Applications. *Batteries* **2019**, *5*, 52. [CrossRef]
4. Sao, C.K.; Lehn, P.W. Control and power management of converter fed micro-grids. *IEEE Trans. Power Syst.* **2008**, *23*, 1088–1098. [CrossRef]
5. Jiayi, H.; Chuanwen, J.; Rong, X. A review on distributed energy resources and MicroGrid. *Renew. Sustain. Energy Rev.* **2008**, *12*, 2472–2483. [CrossRef]
6. Tolba, M.; Rezk, H.; Diab, A.A.; Al-Dhaifallah, M. A novel robust methodology based Salp swarm algorithm for allocation and capacity of renewable distributed generators on distribution grids. *Energies* **2018**, *11*, 2556. [CrossRef]
7. Zeh, A.; Müller, M.; Naumann, M.; Hesse, H.; Jossen, A.; Witzmann, R. Fundamentals of using battery energy storage systems to provide primary control reserves in Germany. *Batteries* **2016**, *2*, 29. [CrossRef]
8. Majumder, R.; Ghosh, A.; Ledwich, G.; Zare, F. Power management and power flow control with back-to-back converters in a utility connected microgrid. *IEEE Trans. Power Syst.* **2010**, *25*, 821–834. [CrossRef]
9. Rezk, H.; Sayed, E.T.; Al-Dhaifallah, M.; Obaid, M. Fuel cell as an effective energy storage in reverse osmosis desalination plant powered by photovoltaic system. *Energy* **2019**, *175*, 423–433. [CrossRef]
10. Rezk, H. A comprehensive sizing methodology for stand-alone battery-less photovoltaic water pumping system under the Egyptian climate. *Cogent Eng.* **2016**, *3*, 1242110. [CrossRef]
11. Pelegov, D.; José, P. Main Drivers of Battery Industry Changes: Electric Vehicles—A Market Overview. *Batteries* **2018**, *4*, 65. [CrossRef]
12. Fathy, A.; Rezk, H.; Nassef, A.M. Robust Hydrogen-Consumption-Minimization Strategy Based Salp Swarm Algorithm for Energy Management of Fuel cell/Supercapacitor/Batteries in Highly Fluctuated Load Condition. *Renew. Energy* **2019**, *139*, 147–160. [CrossRef]
13. Lonkar, M.; Ponnaluri, S. An overview of DC microgrid operation and control. In Proceedings of the IEEE IREC2015 the Sixth International Renewable Energy Congress, Sousse, Tunisia, 24–26 March 2015; pp. 1–6.
14. Ahshan, R.; Iqbal, M.T.; Mann, G.K.; Quaicoe, J.E. Micro-grid system based on renewable power generation units. In Proceedings of the IEEE CCECE 2010, Calgary, AB, Canada, 2–5 May 2010; pp. 1–4.
15. Phurailatpam, C.; Sangral, R.; Rajpurohit, B.S.; Singh, S.N.; Longatt, F.G. Design and analysis of a dc microgrid with centralized battery energy storage system. In Proceedings of the 2015 Annual IEEE India Conference (INDICON), New Delhi, India, 17–20 December 2015; pp. 1–6.
16. Elsied, M.; Oukaour, A.; Gualous, H.; Lo Brutto, O.A. Optimal economic and environment operation of micro-grid power systems. *Energy Convers. Manag.* **2016**, *122*, 182–194. [CrossRef]
17. Vega-Garita, V.; Ramirez Elizondo, L.; Bauer, P.; Sofyan, M.F.; Narayan, N. Energy Management System for the Photovoltaic Battery-Integrated Module. *Energies* **2018**, *11*, 3371. [CrossRef]

18. Eghtedarpour, N.; Farjah, E. Distributed charge/discharge control of energy storages in a renewable-energy-based DC micro-grid. *IET Renew. Power Gener.* **2014**, *8*, 45–57. [CrossRef]
19. Shehata, E.G.; Gaber, M.S.; Ahmed, K.A.; Salama, G.M. Implementation of an energy management algorithm in DC MGs using multi-agent system. *Int. Trans. Electr. Energy Syst.* **2019**, *29*, e2790. [CrossRef]
20. Hadj-Said, S.; Colin, G.; Ketfi-Cherif, A.; Chamaillard, Y. Analytical solution for energy management of parallel hybrid electric vehicles. *IFAC-PapersOnLine* **2017**, *50*, 13872–13877. [CrossRef]
21. Sedaghati, R.; Shakarami, M.R. A novel control strategy and power management of hybrid PV/FC/SC/battery renewable power system-based grid-connected microgrid. *Sustain. Cities Soc.* **2019**, *44*, 830–843. [CrossRef]
22. Herisanu, N.; Marinca, V.; Madescu, G.; Dragan, F. Dynamic response of a permanent magnet synchronous generator to a wind gust. *Energies* **2019**, *12*, 915. [CrossRef]
23. Chaib, A.; Djalloul, A.; Mohamed, K. Control of a solar PV/wind hybrid energy system. *Energy Procedia* **2016**, *95*, 89–97. [CrossRef]
24. Chirkin, V.G.; Khripach, N.A.; Petrichenko, D.A.; Papkin, B.A. A review Of Battery-Supercapacitor Hybrid Energy Storage System Schemes for Power Systems Applications. *Int. J. Mech. Eng. Technol. (IJMET)* **2017**, *8*, 699–707.
25. Nowdeh, S.A.; Chitsaz, M.; Shojaei, H. Designing of Stand-alone PV/FC Hybrid System for Sale Electrisity. *IJECCT* **2013**, *7*, 41–47.
26. Aldo, S. Solar thermochemical production of hydrogen—A review. *Solar Energy* **2015**, *78*, 603–615.
27. Reddy, D.; Barendse, P.; Khan, M. Power electronic interface for combined heat and power systems using high temperature PEM fuel cell technology. In Proceedings of the Power Engineering Society Conference and Exposition in Africa (PowerAfrica), Johannesburg, South Africa, 9–13 July 2012.
28. Sikkabut, S.; Mungporn, P.; Ekkaravarodome, C.; Bizon, N.; Tricoli, P.; Nahid-Mobarakeh, B.; Pierfederici, S.; Davat, B.; Thounthong, P. Control of high-energy high-power densities storage devices by Li-ion battery and supercapacitor for fuel cell/photovoltaic hybrid power plant for autonomous system applications. *IEEE Trans. Ind. Appl.* **2016**, *52*, 4395–4407. [CrossRef]
29. Biswas, I.; Bajpai, P. Control of PV-FC-Battery-SC hybrid system for standalone DC load. In Proceedings of the 2014 IEEE Eighteenth National Power Systems Conference (NPSC), Guwahati, India, 18–20 December 2014; pp. 1–6.
30. Samson, G.T.; Undeland, T.M.; Ulleberg, O.; Vie, P.J. Optimal load sharing strategy in a hybrid power system based on pv/fuel cell/battery/supercapacitor. In Proceedings of the 2009 IEEE International Conference on Clean Electrical Power, Capri, Italy, 9–11 June 2009; pp. 141–146.
31. Njoya Motapon, S. Design and Simulation of a Fuel Cell Hybrid Emergency Power System for A More Electric Aircraft: Evaluation of Energy Management Schemes. Ph.D. Thesis, École de technologie supérieure, Montréal, QC, Canada, 2013.
32. Motapon, S.N.; Dessaint, L.A.; Al-Haddad, K. A Robust H2 Consumption-Minimization-Based Energy Management Strategy for a Fuel Cell Hybrid Emergency Power System of More Electric Aircraft. *IEEE Trans. Ind. Electron.* **2014**, *61*, 6148–6156. [CrossRef]
33. Al-Dhaifallah, M.; Nassef, A.M.; Rezk, H.; Nisar, K.S. Optimal parameter design of fractional order control based INC-MPPT for PV system. *Solar Energy* **2018**, *159*, 650–664. [CrossRef]
34. Rezk, H.; Aly, M.; Al-Dhaifallah, M.; Shoyama, M. Design and Hardware Implementation of New Adaptive Fuzzy Logic-Based MPPT Control Method for Photovoltaic Applications. *IEEE Access* **2019**, *7*, 106427–106438. [CrossRef]
35. Park, J.; Kim, H.; Cho, Y.; Shin, C. Simple Modeling and Simulation of Photovoltaic Panels Using Matlab/Simulink. *Adv. Sci. Technol. Lett.* **2014**, *73*, 147–155.
36. Bellia, H.; Youcef, R.; Fatima, M. A detailed modeling of photovoltaic module using MATLAB. *NRIAG J. Astron. Geophys.* **2014**, *3*, 53–61. [CrossRef]
37. Haberlin, H. Chapter 3, Solar Cells: Their Design Engineering and Operating Principles. In *Photovoltaic System Design and Practice*; Wiley: Hoboken, NJ, USA, 2012; pp. 79–125.
38. Bouraiou, A.; Hamouda, M.; Chaker, A.; Sadok, M.; Mostefaoui, M.; Lachtar, S. Modeling and simulation of Photovoltaic module and array based on one and two diode model using Matlab/Simulink. International Conference on Technologies and Materials for Renewable Energy, Environment and Sustainability, TMREES15. *Energy Procedia* **2015**, *74*, 864–877. [CrossRef]

39. Diab, A.A.Z.; Rezk, H. Global MPPT based on flower pollination and differential evolution algorithms to mitigate partial shading in building integrated PV system. *Solar Energy* **2017**, *157*, 171–186. [CrossRef]

40. Rezk, H.; Fathy, A. Simulation of global MPPT based on teaching–learning-based optimization technique for partially shaded PV system. *Electr. Eng.* **2017**, *99*, 847–859. [CrossRef]

41. Rezk, H.; Fathy, A. A Novel Methodology for Simulating Maximum Power Point Trackers Using Mine Blast Optimization and Teaching Learning Based Optimization Algorithms for Partially Shaded PV System. *J. Renew. Sustain. Energy* **2016**, *8*, 023503.

42. Karami, N.; Moubayed, N.; Outbib, R. General review and classification of different MPPT Techniques. *Renew. Sustain. Energy Rev.* **2017**, *68*, 1–18. [CrossRef]

43. Rezk, H.; Hasaneen, E.-S. A new MATLAB/Simulink model of triple-junction solar cell and MPPT based on artificial neural networks for photovoltaic energy systems. *Ain Shams Eng. J.* **2015**, *6*, 873–881. [CrossRef]

44. Rezk, H.; Eltamaly, A.M. A comprehensive comparison of different MPPT techniques for photovoltaic systems. *Solar Energy* **2015**, *112*, 1–11. [CrossRef]

45. Fathy, A.; Rezk, H. Multi-Verse Optimizer for Identifying the Optimal Parameters of PEMFC Model. *Energy* **2018**, *143*, 634–644. [CrossRef]

46. Rezk, H. Performance of incremental resistance MPPT based proton exchange membrane fuel cell power system. In Proceedings of the 2016 Eighteenth International Middle East Power Systems Conference (MEPCON), Cairo, Egypt, 27–29 December 2016.

47. Larminie, J.; Dicks, A. *Fuel Cell Systems Explained*, 2nd ed.; John Wiley & Sons Ltd.: Hoboken, NJ, USA, 2003; pp. 1–418.

48. Gebregergis, A.; Pillay, P.; Rengaswamy, R. PEMFC fault diagnosis, modeling and mitigation. *Trans. Ind. Appl.* **2010**, *46*, 295–303. [CrossRef]

49. Mwinga, M.; Groenewald, B.; McPherson, M. Design, Modelling, And Simulation of a Fuel Cell Power Conditioning system. *J. Therm. Eng.* **2015**, *1*, 408–419. [CrossRef]

50. Tremblay, O.; Louis-A, D. A generic fuel cell model for the simulation of fuel cell vehicles. In Proceedings of the 2009 IEEE Vehicle Power and Propulsion Conference, Dearborn, MI, USA, 7–10 September 2009.

51. Outeiro, M. MatLab/Simulink as design tool of PEM Fuel Cells as electrical generation systems. *Eur. Fuel Cell Forum* **2011**, *28*, 1–9.

52. Bajpai, P.; Dash, V. Hybrid renewable energy systems for power generation in stand-alone applications: A review. *Renew. Sustain. Energy Rev.* **2012**, *16*, 2926–2939. [CrossRef]

53. Boulmrharj, S.; NaitMalek, Y.; Elmouatamid, A.; Bakhouya, M.; Ouladsine, R.; Zine-Dine, K.; Khaidar, M.; Siniti, M. Battery Characterization and Dimensioning Approaches for Micro-Grid Systems. *Energies* **2019**, *12*, 1305. [CrossRef]

54. Fotouhi, A.; Auger, D.J.; Propp, K.; Longo, S.; Wild, M. A review on electric vehicle battery modelling: From Lithium-ion toward Lithium–Sulphur. *J Renew. Sustain. Energy Rev.* **2016**, *56*, 1008–1021. [CrossRef]

55. Quinard, H.; Redondo-Iglesias, E.; Pelissier, S.; Venet, P. Fast Electrical Characterizations of High-Energy Second Life Lithium-Ion Batteries for Embedded and Stationary Applications. *Batteries* **2019**, *5*, 33. [CrossRef]

56. Madani, S.S.; Schaltz, E.; Knudsen Kær, S. An Electrical Equivalent Circuit Model of a Lithium Titanate Oxide Battery. *Batteries*. **2019**, *5*, 31. [CrossRef]

57. Jiang, J.; Liang, Y.; Ju, Q.; Zhang, L.; Zhang, W.; Zhang, C. An equivalent circuit model for lithium-sulfur batteries. *Energy Procedia* **2017**, *105*, 3533–3538. [CrossRef]

58. Valant, C.; Gaustad, G.; Nenadic, N. Characterizing Large-Scale, Electric-Vehicle Lithium Ion Transportation Batteries for Secondary Uses in Grid Applications. *Batteries* **2019**, *5*, 8. [CrossRef]

59. Saw, L.H.; Somasundaram, K.; Ye, Y.; Tay, A.A. Electro-thermal analysis of Lithium Iron Phosphate battery for electric vehicles. *J. Power Sources* **2014**, *249*, 231–238. [CrossRef]

60. Donghwa, S.; Kim, Y.; Seo, J.; Chang, N.; Wang, Y.; Pedram, M. Battery-supercapacitor hybrid system for high-rate pulsed load applications. In Proceedings of the Design, Automation & Test in Europe, Grenoble, France, 14–18 March 2011; pp. 1–4.

61. Van Mierlo, J.; Van den Bossche, P.; Maggetto, G. Models of energy sources for EV and HEV: Fuel cells, batteries, ultracapacitors, flywheels and engine-generators. *J. Power Sources* **2003**, *128*, 76–89. [CrossRef]

62. Uzunoglu, M.; Onar, O.C.; Alam, M.S. Modeling, control and simulation of a PV/FC/UC based hybrid power generation system for stand-alone applications. *Renew. Energy* **2009**, *34*, 509–520. [CrossRef]

63. Rullo, P.; Braccia, L.; Luppi, P.; Zumoffen, D.; Feroldi, D. Integration of sizing and energy management based on economic predictive control for standalone hybrid renewable energy systems. *Renew. Energy* **2019**, *140*, 436–445. [CrossRef]

64. Merabet, A.; Ahmed, K.T.; Ibrahim, H.; Beguenane, R.; Ghias, A.M. Energy management and control system for laboratory scale microgrid based wind-PV-battery. *IEEE Trans. Sustain. Energy* **2016**, *8*, 145–154. [CrossRef]

65. Behrooz, F.; Mariun, N.; Marhaban, M.H.; Radzi, M.A.M.; Ramli, A.R. Review of control techniques for HVAC systems—Nonlinearity approaches based on Fuzzy cognitive maps. *Energies* **2018**, *11*, 495. [CrossRef]

66. Li, Q.; Chen, W.; Li, Y.; Liu, S.; Huang, J. Energy management strategy for fuel cell/battery/ultracapacitor hybrid vehicle based on fuzzy logic. *Electr. Power Energy Syst.* **2012**, *43*, 514–525. [CrossRef]

67. Liu, G.; Zhang, J.; Sun, Y. High frequency decoupling strategy for the PEM fuel cell hybrid system. *Int. J. Hydrogen Energy* **2008**, *33*, 6253–6261. [CrossRef]

Article

Copper-Decorated CNTs as a Possible Electrode Material in Supercapacitors

Mateusz Ciszewski [1],*, Dawid Janas [2,3] and Krzysztof K. Koziol [3]

[1] Department of Hydrometallurgy, Łukasiewicz Research Network—Institute of Non-Ferrous Metals, Sowińskiego 5, 44-100 Gliwice, Poland
[2] Department of Organic Chemistry, Bioorganic Chemistry and Biotechnology, Silesian University of Technology, B. Krzywoustego 4, 44-100 Gliwice, Poland
[3] Department of Transport and Manufacturing, Cranfield University, College Road, Cranfield, Bedford MK43 0AL, UK
* Correspondence: mateuszc@imn.gliwice.pl; Tel.: +48-32-2380277

Received: 30 May 2019; Accepted: 4 July 2019; Published: 3 September 2019

Abstract: Copper is probably one of the most important metal used in the broad range of electronic applications. It has been developed for many decades, and so it is very hard to make any further advances in its electrical and thermal performance by simply changing the manufacture to even more oxygen-free conditions. Carbon nanotubes (CNTs) due to their excellent electrical, thermal and mechanical properties seem like an ideal component to produce Cu-CNT composites of superior electrochemical performance. In this report we present whether Cu-CNT contact has a beneficial influence for manufacturing of a new type of carbon-based supercapacitor with embedded copper particles. The prepared electrode material was examined in symmetric cell configuration. The specific capacity and cyclability of composite were compared to parent CNT and oxidized CNT.

Keywords: CNT; copper; composite; energy storage

1. Introduction

Carbon nanotubes (CNTs) are quasi-1D nano materials that have attracted significant attention because of remarkable properties such as high chemical stability, high electrical conductivity, strength, flexibility and impressive surface area [1–4]. All these features make CNTs very appealing for various applications such as hydrogen storage, fuel cells, supercapacitors, lithium ion batteries, stealth paints, sensors, gas and toxin detection systems, catalysts, etc. [5–8]. CNTs also may be used as an additive component to enhance properties of various materials [9–12]. Many attempts have been made to deposit metal or metal compounds into and onto CNTs. One of the most obvious routes would be melting of metal onto the CNT surface, however, most metals are not able to wet its surface [13]. Typically three methods are in use to manufacture CNT composites: mechanical alloying of metal nanopowders with CNT in a planetary mill (and further annealing or plasma sintering), molecular level mixing (i.e., CNTs functionalization and reaction with metal salt) and electrodeposition process [14]. The results show that pre-functionalization of CNTs with oxygen-containing functional groups gives much better interaction with the metal matrix [15]. As-made CNTs are sparsely dispersed in every common medium therefore they have to be functionalized before metal incorporation. The simplest methodology is oxidation, which creates oxygen-containing groups being active sites for metal deposition. Many different chemical oxidation techniques of CNTs have been proposed using concentrated mineral acids additionally enhanced by oxidizing agent for example HNO_3, H_2SO_4, aqua regia, $KMnO_4$, $OsO_4/NaIO_4$ [16–21]. Nitric acid or combination of nitric acid and sulfuric acid are most commonly used to activate CNT surface. Here important seems to be a proper ratio of the oxidative acid mixture, which is responsible for attaching oxygen species. The disadvantage is that such approach ultimately leads to disruption of the highly-conductive sp^2 network of carbon atoms if the degree of functionalization is too high [22]. Although oxidation of CNTs often suffer from low precision and poor control of the final material, it is the simplest and most conventional pretreatment method for preparation of high-performance CNT-metal composites.

Although Cu-CNT composites have attracted much attention recently [23,24], it should be pointed out that despite its high electrical conductivity Cu produces only a weak bonding with carbon matrix and combining these two material is a challenge [15]. To achieve low-resistance ohmic contacts, and thus create a Cu-CNT composite of appreciable properties, improvement in connection between CNT and copper at the nanoscale is crucial. This is commonly handled by manufacture of CNT—copper composites by wet preparation methods, which are composed of multi component single-step or two step sensitization—activation processes in tin chloride, palladium chloride, copper salt, sodium formate, formaldehyde, ethylenediaminetetraacetic acid (EDTA) and polyethyleneglycole-containing bath [25,26]. Here we report facile preparation of Cu-CNT composites by CNTs acid functionalization and fast copper salt reduction. To confirm that this material may have appreciable electrical properties, we evaluated its potential for the application in supercapacitors. The product was configured in two-electrode assembly to probe its electrochemical performance.

2. Experimental

2.1. CNT Synthesis and Oxidation

Vertically-aligned carbon nanotubes were obtained in a chemical vapor deposition process from toluene (HPLC grade, Fisher Scientific, Hempton, NH, USA) as a carbon precursor and ferrocene (99.5%, Alfa Aesar, Haverhill, MA, USA) as a catalyst [27]. Toluene was ultrasonicated with 3 wt. % ferrocene just before the process. Reaction was carried out in a horizontal furnace with three heating sections under argon atmosphere at 760 °C. After the synthesis, 0.5 g of freshly prepared CNT were mixed with 187.5 mL H_2SO_4 (>95%, Fisher Scientific) and 62.5 mL HNO_3 (70%, Fisher Scientific) in a 500 mL flask, and then ultrasonicated at 50 °C for 20 h. Next, the material was diluted with deionized water to 2 L and filtrated using polytetrafluoroethylene (PTFE) membrane filters (pore size 0.45 μm, Whatman, Maidstone, UK) on a fritted-glass funnel under vacuum and dried overnight at 110 °C. The oxidized CNTs were denoted in further experiments and analysis as CNTOX.

2.2. Preparation of Cu-CNT Composites

20 mg of as-oxidized CNTs was ultrasonically re-dispersed in 10 mL of distilled water, and then suspension was combined with metal precursor copper (II) sulfate pentahydrate (Fisher Scientific, Hempton, NH, USA). The mixture was ultrasonicated overnight. It was subsequently treated with 2 mL hydrazine hydrate (78–82%, Sigma-Aldrich, St Louis, MO, USA). Color change from black to brown was immediately observed, which indicated formation of metallic copper. Product was filtrated on PTFE membrane filter and obtained slurry (denoted as Cu-CNT) was put into the oven (120 °C) to remove the solvent.

2.3. Characterization and Electrochemical Testing

The materials were qualitatively analyzed using infra-red spectrometer (Bruker Optics, IFS 66/s Billerica, MA, USA) and Raman spectroscope (Renishaw, inVia, λ = 633 nm laser Wotton under Edge, UK). Carbon amount in a primary material and oxidized product, copper load in composites and impurities content were quantitatively evaluated by EDX analysis (EDX, Bruker Quantax coupled with Nova NanoSEM, Goettingem, Germany). All materials were additionally tested by thermogravimetry and calorimetry (Mettler Toledo, TGA/DSC Columbus, OH, USA). Morphology was examined by means of Scanning Electron Microscopy (SEM, FEI Nova NanoSEM). Electrochemical experiments were carried out using two-electrode system. The working electrode materials were placed on electrochemical nickel current collectors to form films and separated with membrane (Whatman) soaked with 1 M Na_2SO_4. Electrodes, current collectors and separator were pressed with four screws in a poly (methyl methacrylate) casing. Cyclic voltammetry (CV) and galvanostatic charge/discharge (GC) characteristics were performed with Autolab PGSTAT 30 workstation.

3. Results

Microstructure of neat CNTs and oxidized form (CNTOX) were analyzed using SEM to find changes in structure produced during acid treatment. As-made material was composed of vertically-aligned CNTs tens of μm long (Figure 1a,b). EDX study confirmed about 98 wt% carbon and residual Fe catalyst (2 wt%) (Figure 2a).

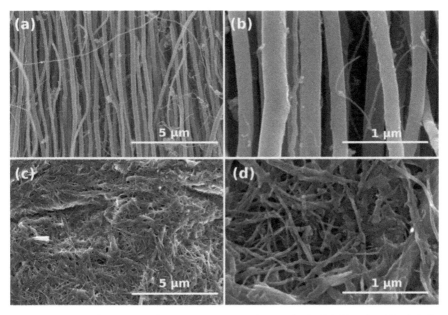

Figure 1. Low-and high-magnification SEM images of as-synthesized (**a,b**) and oxidized (**c,d**) CNTs, respectively.

Acid treatment induced defects, shortened CNTs and increased densification degree of the material (Figure 1c,d). Moreover, the entangled CNTOX material was no longer anisotropic. Presence of the oxygen in CNTOX was confirmed by EDX analysis as shown in Figure 2b.

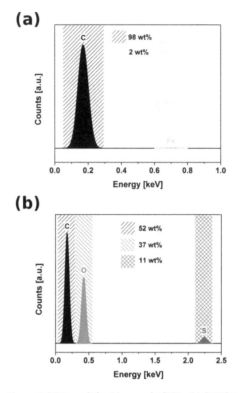

Figure 2. EDX result for (a) as-made CNTs, (b) CNTOX.

Acid treatment introduced ca. 37 wt% oxygen. Additionally, 11 wt% of sulfur, which is caused by contamination of oxidized CNT with strongly adsorbed sulfur-based chemical groups (most probably sulfate) introduced by sulfuric acid (could not be removed by the work-up despite copious amounts of distilled water used for filtration), was detected.

SEM images of Cu-CNT composite revealed presence of copper in the carbon matrix (Figure 3). CNTs were covered by two types of particles, which were equally distributed: very fine Cu particles (Figure 3b) and regions with bigger agglomerates (Figure 3c) containing copper, sulfur and oxygen (probably unreduced copper sulfate or some copper oxide species).

EDX analysis showed 79 wt% Cu and 13 wt% C with the rest attributed to the residual oxygen groups from CNT functionalization and Cu precursor (Figure 4a).

Raman spectroscopy evaluated changes in the chemical structure of the material in the course of functionalization and Cu deposition (Figure 4b). The higher the ratio of the D to G band intensity the more disordered structure of the analyzed material. The D band located around 1350 cm^{-1} is a consequence of structure disorder by the introduction of sp^3 hybridized carbon atoms, whereas G band around 1600 cm^{-1} determines carbon domain with sp^2 hybridization. Ordered structure of as-synthesized CNTs had relatively low D/G ratio around 0.36. Acid treated CNTs had plethora of oxygen-bearing functional groups and were partially unzipped, much more corrugated and consequently more disordered. This resulted in an increase in D band intensity to 0.92. Oxidized CNT matrix coated with copper precursor and reduced using hydrazine was even more affected because of chemical reduction which caused elimination of some carbon atoms from CNT structure together with the oxygen. Consequently, the D/G ratio for the Cu-CNTOX composite was as high as 0.99.

Figure 3. SEM images at various magnifications of Cu-CNT composites with two types of particles decorated onto carbon matrix.

Qualitative analysis was performed using FTIR (Figure 5). In comparison to the as-made CNT the oxidized form had several signals with three of them most important: at 1750 cm^{-1} assigned to the carboxylic groups, 1620 cm^{-1} to hydroxyl groups and around 1220 cm^{-1} probably from sulfates embedded during acid treatment. Because of reduction of copper precursor and oxidized CNT matrix by hydrazine hydrate most of signals vanished in Cu-CNT. Only presence of the residual hydroxyl groups can be observed. Additionally signals located at 2200–2000 cm^{-1} ascribed to carbon structure were present in all three spectra, however, in the Cu-CNT composite they had the smallest intensities probably because of high copper loading. Not observable in this figure because of spectra overlapping was a broad and small signal located at 3500–3000 cm^{-1} from stretching vibration of C–OH groups and water.

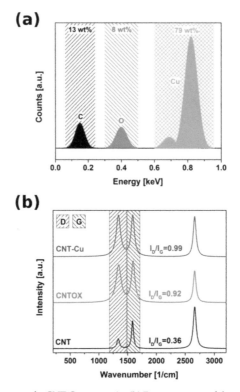

Figure 4. (**a**) EDX spectrum of a CNT-Cu composite (**b**) Raman spectra of the as-made CNT, CNTOX and CNT-Cu composite.

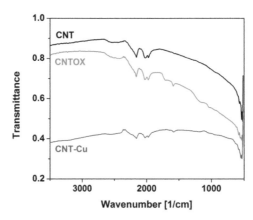

Figure 5. FTIR spectra of the as-made CNT, CNTOX and CNT-Cu composite.

TGA and DSC curves recorded for all tested materials were presented in Figure 6. In CNT only one gravimetric change can be found that was assigned to combustion of carbon honeycomb matrix that was satisfied by the presence of intensive exothermic signal in DSC curve. For the CNTOX matrix thermogravimetric curve was composed of several steps related to removal of residual solvent around 100 °C, water of hydration starting from 220 °C, gradual elimination of the oxygen containing groups up to 540 °C and decomposition of carbon matrix close to 640 °C. In both CNT and CNTOX weight loss was 100%, which satisfied absence of any significant metal impurities, with a strong exothermic signal attributed to carbon combustion. In the copper—CNT composite carbon matrix weight loss was observed at lower temperature than in CNT and CNTOX that could have been caused by catalytic effect of copper. In the temperature range 410 °C–700 °C there was also oxidation of the residual $CuSO_4$ (left after hydrazine reduction) to copper sulfate hydroxide hydrate and copper sulfate hydroxide, but their DSC signal was much lower in comparison to dominant carbon decomposition. As the amount of carbon with respect to copper in the composite was smaller the TG curve step was also not so significant. Final weight of the sample was about 70–75% of the initial—because carbon loss was equilibrated by copper oxidation compounds.

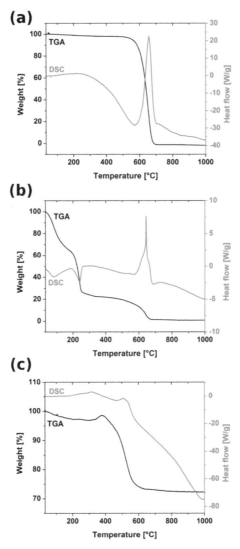

Figure 6. TGA (black) and DSC (red) curves recorded for (**a**) as-made CNT, (**b**) CNTOX and (**c**) Cu-CNT composites.

As obtained Cu-CNT composite was electrochemically tested using two electrode symmetric cell as a potential material for supercapacitor electrodes. Results were compared to the as-made CNT as well as CNTOX. Figure 7a presents CV curves obtained for tested materials during 1000 charge discharge cycles at scan rate 500 mV/s in the potential window 0–1 V. In case of CNT curves were overlaid with the *x*-axis as the current intensity was very low of the order of 0.00001 A. In contrary both CNTOX and Cu-CNT curves had much higher value of current intensity ca. 0.001 A. The more ideal box-like shape was obtained for Cu-CNT whereas for CNTOX more asymmetric curves with a strong oxidizing peak near 0.5V, particularly in the first few cycles, were registered. That was caused by irreversible redox reactions of electrode material with an electrolyte. Capacity loss after 1000 cycles was as high as 5% for CNT, 30.4% for CNTOX and 16% for Cu-CNT. However, it should be emphasized that the difference between first and last step in Cu-CNT was as high as 48% (different shape for three initial cycles), which is a consequence of chemical reaction of electrolyte and active species.

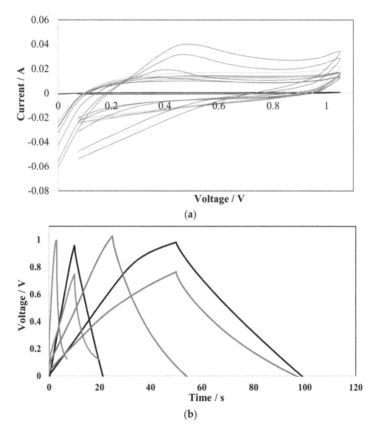

Figure 7. (a) CV curves and (b) galvanostatic charge discharge curves recorded for CNT (black), CNTOX (red) and Cu-CNT (blue).

The real specific capacity was calculated from the galvanostatic charge/discharge curves (Figure 7b). Curves recorded for all three types of materials were very similar, however, the least symmetric shape was found for copper-based composite. This was caused by reactions of copper, oxygen and residual sulfur present within material with an electrolyte. Consequently, the concept of Cu incorporation onto CNT may be a proper route to enhance electrical and capacity properties of CNT, but more efforts should be done to obtain material of high purity that will limit effect of undesirable side reactions. In case of sparsely conductive CNTOX, relatively high specific capacity and shape of galvanostatic curves was obtained. The oxygen containing groups are responsible for pseudocapacitance that enhances specific capacity, however, very often these species cause self-discharge of electrode material. Measurements were performed in a broad range of current density from 0.02 A/g to 7 A/g. The calculated specific capacity was 1.2 F/g, 25 F/g and 46 F/g for CNT, CNTOX and Cu-CNT, respectively.

It was relatively easy to charge and discharge CNT with resulting in proper triangular shape of galvanostatic curve, similarly for CNTOX, while for Cu-CNT problems with complete discharge were found. This showed that incorporation of copper copper-bearing species strongly limited pores accessibility and induced some irreversible redox reactions. Although materials morphology was changed the net specific capacity of the composite was enhanced based on pseudocapacitive effect.

4. Conclusions

It was demonstrated that incorporation of copper metal into CNT matrix may be done using simple solvothermal wet impregnation method. CNT matrix was oxidatively functionalized using acid treatment. Produced acid active sites were used to anchor metal seeds, which were finally chemically reduced. This showed that copper can be effectively combined with CNT in a simple process. Analysis estimated ca. 70 wt% copper in the composite. Metal was equally distributed as fine particles with some regions of bigger agglomerates of unreacted copper precursor. As prepared material was examined as the electrode for supercapacitors. The results showed that such Cu-CNT composite can be used in energy storage materials but the active species, because of its high activity, has to be purified. Further advances in creating such material with better contact between Cu and CNTs, lack of agglomerates and lower degree of impurities could give much higher performance. Eliminating these issues would enable much better electron transport through the material. What regards CNTs themselves, gaining control over chirality and employing SWCNTs could show us what is the ultimate limit of performance in terms of supercapacitance, electrical conductivity and other applications of Cu-CNT composites.

Author Contributions: Conceptualization, M.C.; methodology, M.C.; formal analysis, M.C.; investigation, M.C.; resources, M.C.; data curation, D.J.; writing—original draft preparation, M.C.; writing—review and editing, D.J.; visualization, M.C. and D.J.; supervision, K.K.K.; project administration, K.K.K.

Funding: This research received no external funding.

Conflicts of Interest: The authors declare no conflict of interest.

References

1. Merkoci, A.; Pumera, M.; Llopis, X.; Perez, B.; del Valle, M.; Alegret, S. New materials for electrochemical sensing VI: Carbon nanotubes. *TrAC-trend. Anal. Chem.* **2005**, *24*, 826–838. [CrossRef]

2. Ruoff, R.S.; Qian, D.; Liu, W.K. Mechanical properties of carbon nanotubes: theoretical predictions and experimental measurements. *Physics* **2003**, *4*, 993–1008. [CrossRef]

3. Peigney, A.; Laurent, C.; Flahaut, E.; Bacsa, R.R.; Rousset, A. Specific surface area of carbon nanotubes and bundles of carbon nanotubes. *Carbon* **2001**, *39*, 507–514. [CrossRef]

4. Hone, J.; Llaguno, M.C.; Biercuk, M.J.; Johnson, A.T.; Batlogg, B.; Benes, Z.; Fischer, J.E. Thermal properties of carbon nanotubes and nanotube-based materials. *Appl. Phys. A* **2002**, *74*, 339–343. [CrossRef]

5. De Volder, M.F.; Tawfick, S.H.; Baughman, R.H.; Hart, A.J. Carbon nanotubes: present and future commercial applications. *Science* **2013**, *339*, 535–539. [CrossRef] [PubMed]

6. Sehrawat, P.; Julien, C.; Islam, S.S. Carbon nanotubes in Li-ion batteries: A review. *Mater. Sci. Eng. B* **2016**, *213*, 12–40. [CrossRef]

7. Rakov, E.G.; Baronin, I.V.; Anoshkin, I.V. Carbon nanotubes for catalytic applications. *Catal. Ind.* **2010**, *2*, 26–28. [CrossRef]

8. Aroutiounian, V.M. Gas sensors based on functionalized carbon nanotubes. *J. Contemp. Phys.* **2015**, *50*, 333–354. [CrossRef]

9. Saeed, S.; Hakeem, S.; Faheem, M.; Alvi, R.A.; Farooq, K.; Hussain, S.T.; Ahmad, S.N. Effect of doping of multi-walled carbon nanotubes on phenolic based carbon fiber reinforced nanocomposites. *J. Phys. Conf. Ser.* **2013**, *439*, 012017. [CrossRef]

10. Lee, R.H.; Lee, L.Y.; Huang, J.L.; Huang, C.C.; Hwang, J.C. Conjugated polymer-functionalized carbon nanotubes enhance the photovoltaic properties of polymer solar cells. *Colloid. Polym. Sci.* **2011**, *289*, 1633–1641. [CrossRef]

11. Zhou, C.; Li, F.; Hu, J.; Ren, M.; Wei, J.; Yu, Q. Enhanced mechanical properties of cement paste by hybrid graphene oxide/carbon nanotubes. *Constr. Build. Mater.* **2017**, *134*, 336–345. [CrossRef]

12. Zheng, J.; Li, M.; Yu, K.; Hu, J.; Zhang, X.; Wang, L. Sulfonated multiwall carbon nanotubes assisted thin-film nanocomposite membrane with enhanced water flux and anti-fouling property. *J. Memb. Sci.* **2017**, *524*, 344–353. [CrossRef]

13. Ebbesen, T.W. Wetting, filling and decorating carbon nanotubes. *J. Phys. Chem. Solids* **1996**, *57*, 951–955. [CrossRef]

14. Tjong, S.C. *Carbon Nanotube Reinforced Composites: Metal and Ceramic Matrices*; WILEY-VCH Verlag GmbH & Co. KgaA: Weinheim, Germany, 2009.

15. Park, M.; Kim, B.H.; Kim, S.; Han, D.S.; Kim, G.; Lee, K.R. Improved binding between copper and carbon nanotubes in a composite using oxygen-containing functional groups. *Carbon* **2011**, *49*, 811–818. [CrossRef]

16. Satishkumar, B.C.; Govindaraj, A.; Mofokeng, J.; Subbanna, G.N.; Rao, C.N.R. Novel experiments with carbon nanotubes: opening, filling, closing and functionalizing nanotubes. *J. Phys. B A. Mol. Opt. Phys.* **1996**, *29*, 4925–4934. [CrossRef]

17. Tsang, S.C.; Chen, Y.K.; Harris, P.J.F.; Green, M.L.H. A simple chemical method of opening and filling carbon nanotubes. *Nature* **1994**, *372*, 159–162. [CrossRef]

18. Colomer, J.F.; Piedigrosso, P.; Fonseca, A.; Nagy, J.B. Different purification methods of carbon nanotubes produced by catalytic synthesis. *Synth. Met.* **1999**, *103*, 2482–2483. [CrossRef]

19. Hernadi, K.; Siska, A.; Thien-Nga, L.; Forro, L.; Kiricsi, I. Reactivity of different kinds of carbon during oxidative purification of catalytically prepared carbon nanotubes. *Solid State Ionics* **2001**, *141–142*, 203–209. [CrossRef]

20. Rosca, I.D.; Watari, F.; Uo, M.; Akasaka, T. Oxidation of multiwalled carbon nanotubes by nitric acid. *Carbon* **2005**, *43*, 3124–3131. [CrossRef]

21. Datsyuk, V.; Kalyva, M.; Papagelis, K.; Parthenios, J.; Tasis, D.; Siokou, A.; Kallitsis, I.; Galiotis, C. Chemical oxidation of multiwalled carbon nanotubes. *Carbon* **2008**, *46*, 833–840. [CrossRef]

22. Hu, H.; Zhao, B.; Itkis, M.E.; Haddon, R.C. Nitric Acid Purification of Single-Walled Carbon Nanotubes. *J. Phys. Chem. B* **2003**, *107*, 13838–13842.

23. Huan, H.; Fu, B.; Ye, X. The torsional mechanical properties of copper nanowires supported by carbon nanotubes. *Phys. Lett. A* **2017**, *381*, 481–488. [CrossRef]

24. Gamat, S.N.; Fotouhi, L.; Talebpour, Z. The application of electrochemical detection in capillary electrophoresis. *J. Iran. Chem. Soc.* **2017**, *14*, 717–725. [CrossRef]

25. Xu, C.; Wu, G.; Liu, Z.; Wu, D.; Meek, T.T.; Han, Q. Preparation of copper nanoparticles on carbon nanotubes by electroless plating, Materials Research Bulletin. *Mater. Res. Bull.* **2004**, *39*, 1499–1505. [CrossRef]

26. Ang, L.M.; Hor, T.S.A.; Xu, G.Q.; Tung, C.H.; Zhao, S.P.; Wang, J.L.S. Decoration of activated carbon nanotubes with copper and nickel. *Carbon* **2000**, *38*, 363–372. [CrossRef]

27. Pattinson, S.W.; Prehn, K.; Kinloch, I.A.; Eder, D.; Koziol, K.K.K.; Schulte, K.; Windle, A.H. The life and death of carbon nanotubes. *RSC Adv.* **2012**, *2*, 2909–2913. [CrossRef]

 *batteries*

Case Report

From Bench-Scale to Prototype: Case Study on a Nickel Hydroxide—Activated Carbon Hybrid Energy Storage Device

Alberto Adan-Mas [1,*], Pablo Arévalo-Cid [1], Teresa Moura e Silva [2], João Crespo [3] and Maria de Fatima Montemor [1]

[1] Centro de Química Estrutural-CQE, DEQ, Instituto Superior Tecnico, Universidade de Lisboa, 1049-001 Lisboa, Portugal; pabloarevalo@quim.ucm.es (P.A.-C.); mfmontemor@tecnico.ulisboa.pt (M.d.F.M.)
[2] ADEM, ISEL-Instituto Superior de Engenharia de Lisboa, Instituto Politécnico de Lisboa, 1959-007 Lisboa, Portugal; msilva@dem.isel.ipl.pt
[3] Arquiled Projectos de Iluminação S.A., Edifício Arcis, R. Ivone Silva 6, 17Esq, 1050-124 Lisboa, Portugal; joao.crespo@arquiled.com
* Correspondence: alberto.mas@tecnico.ulisboa.pt

Received: 14 September 2019; Accepted: 8 October 2019; Published: 15 October 2019

Abstract: Hybrid capacitors have been developed to bridge the gap between batteries and ultracapacitors. These devices combine a capacitive electrode and a battery-like material to achieve high energy-density high power-density devices with good cycling stability. In the quest of improved electrochemical responses, several hybrid devices have been proposed. However, they are usually limited to bench-scale prototypes that would likely face severe challenges during a scaling up process. The present case study reports the production of a hybrid prototype consisting of commercial activated carbon and nickel-cobalt hydroxide, obtained by chemical co-precipitation, separated by means of polyolefin-based paper. Developed to power a 12 W LED light, these materials were assembled and characterized in a coin-cell configuration and stacked to increase device voltage. All the processes have been adapted and constrained to scalable conditions to ensure reliable production of a pre-commercial device. Important challenges and limitations of this process, from geometrical constraints to increased resistance, are reported alongside their impact and optimization on the final performance, stability, and metrics of the assembled prototype.

Keywords: nickel-cobalt hydroxide; activated carbon; hybrid capacitor prototype case study; KOH aqueous electrolyte energy storage device; coin-cell prototype

---

1. Introduction

Understanding technological growth and development is vital for societal progress. The technology life-cycle (TLC) defines the timeline of a manufacturing process, from its conception to culmination. Its first major step is the "research and development phase", which entails high risk of failure and that shall be completed before moving to an ascendant phase of commercialization [1]. To evaluate state-of-progress during the R&D phase, another tool, known as the technology readiness level (TRL) scale, is introduced.

Developed by the National Aeronautics and Space Administration (NASA), the TRL scale is currently employed to evaluate the maturity of a given technology during an innovation process, summarizing its risks and opportunities. In recent years, the TRL scale has been adopted by the European Union to evaluate the funding of project proposals by classifying different research stages in nine progressive levels [2]. These evolve from the observation of basic principles (TRL1) and Technology concept formulated (TRL2) up to "system proven in operational environment" (TRL 9), passing through several steps including fabrication and validation of prototypes in relevant environments.

During the conception of research & development proposals, the objective is to progress from a low TRL to, at least, the following level. For example, experimental science often targets to create a proof of concept that can be later validated in industrial conditions with the goal of transferring the technology from the research institution or facilities to a commercializing process or business. In that way, it is possible to validate the impact and potential application of the project to society.

In the case of research focused on energy storage, recent advances in material science have led to novel energy storage materials and technologies. These can potentially have a major impact on one of the seventeen most important goals for sustainable development, as stated by the United Nations [3], "Ensure access to affordable, reliable, sustainable and modern energy". During the last decade, research has focused intensively in producing novel materials with enhanced properties to improve the current state-of-the art electrochemical energy storage devices; from lithium-batteries to ultracapacitors [4]. Lithium-ion batteries are, to the date, the most efficient and commercialized energy storage devices. Nonetheless, technological developments impose energy storage requirements that Li-ion batteries cannot meet [5]. Alternative technologies are, therefore, under investigation; from lithium-air batteries [6] to aluminum-ion batteries [7]. The present case study reports some of the challenges faced, at an experimental level, during the development of a potentially scalable hybrid capacitor prototype.

Hybrid capacitor devices have been explored to bridge the gap between the high-power capabilities of electrical double-layer capacitors and the high-energy density of batteries. To that end, they integrate a capacitive electrode, usually carbon-based, that ensures high-power density and cycling stability and a battery-like material, based on metal compounds (typically oxides or hydroxides), that increase the overall energy that can be stored in the device. The different performance of the constituent electrodes is a consequence of their different energy storing nature, while the former is based on electrostatic adsorption/desorption of ions at the electrode-electrolyte interface; the latter relies on faradaic processes that store energy by means of chemical reactions [8].

In an effort to scale-up the production of a hybrid energy storage prototype, the present case study reports the assembly of a hybrid device to power a 12 W solar-powered LED lamp, with the consequent challenges and limitations faced. This contribution will guide the readers across the difficult art of developing a bench-scale prototype, starting from the preparation of the material and testing of the device to meeting the requirements of the final application.

2. Device Conceptualization

2.1. Material Selection

The first aspect to consider when building an energy storage prototype is the active material selection for both electrodes, positive and negative, as this will determine the device electrochemical response, device performance stability, shelf life and overall metrics. An adequate choice of materials for the electrodes must consider materials with good electrochemical performance and well-matched properties, active in the same electrolyte and that can be produced in large scale at an affordable price. Concomitantly, the selection of the electrolyte, preferably an environmentally friendly one, is a critical parameter and must be considered in the selection of the active materials. The recent trends highlight a shift towards aqueous and "greener" electrolytes that can replace conventional organic electrolytes that possess certain toxicity and have lower tolerance to high environmental temperature. Altogether, both materials and electrolyte must be synergistically combined. Therefore, for the targeted application, a solar powered LED light, an aqueous electrolyte will lead to less risks of explosion, more material availability, lower cost and environmental sustainability.

A hybrid device requires a battery-like and an electrical double layer capacitive-type (EDLC) electrode. Material production is limited by performance, production rate, and easy scale-up production. The materials selected for this case study are nickel-cobalt hydroxide, synthetized at the laboratory, and commercial activated carbon. While the former ensures the energy metrics, the latter determines the power capabilities. These two materials, apart from fulfilling the imposed electrochemical requirements, have been recently explored in literature as excellent energy storage materials [9]. In fact, they are two of the most investigated materials and have been reported as two of the better performing materials known to the date. In addition, both are active in alkaline media, such as KOH 6M, which is used as electrolyte; they can be easily synthesized by scalable processes; they present acceptable cycling stability and its combination complies with the high-power imposed requirements.

An important aspect when using a laboratory-produced material is the scale-up. To ensure high-yield production, nickel-cobalt hydroxide was produced by co-precipitation. Therefore, the first constraint to be noted is the synthesis being limited by reactor sizes, reaction times and need of additional steps for separation and purification. This issue must be evaluated when planning the construction of a prototype and, it is crucial to consider that scale-up may affect the material properties (e.g., crystalline structure, morphology, composition, etc.), consequently having an impact on the final electrochemical response.

The proposed materials for the electrode assembly are available in powder form. Thus, an ink must be prepared and coated on top of a conductive substrate. Active material inks can be easily prepared under standard conditions [10], this is, 10% conductive carbon (CB), 5% PVDF and 85% active material dispersed in N-Methyl-2-Pyrrolidone (NMP) for nickel-cobalt hydroxide and 5% CB, 5% PVDF and 90% activated carbon dispersed in the same solvent for the negative electrode. When introducing these additives, two important constraints are introduced: on the one hand, non-active mass and, on the other hand, poorly conductive species that may affect the electrode capacity.

A hybrid energy storage device typically targets good energy density and excellent power performance. For that reason, it is important to combine capacitive and faradaic materials that can act synergistically. For that matter, a carbon-based substrate may serve as excellent conductive matrix to enhance faradaic phenomena. Moreover, it ensures high surface roughness, good electrolyte penetrability, and high electronic conductivity and can introduce relevant capacitance contribution while maintaining high-power capabilities. Moreover, it presents low density, a key parameter in the final weight of the device. For these reasons, Toray carbon paper has been selected as conductive substrate for both the positive and negative electrodes. At this point is worth noting that the electronic conductivity and mechanical integrity of these substrates must be ensured over the entire assembling process.

Finally, an adequate separator with chemical stability in the designated electrolyte must be used. To that end, commercially available cellulose-based separators can be used. However, certain separators may entail stability problems under certain conditions. For example, certain polyamide separators may decompose under a highly alkaline media that may release ammoniacal gases. Adequate separator thickness shall also be considered. Excessive thickness may lead to increased resistance and, consequently, lower power and energy density response of the device. For that reason, a 0.12 mm thick polyolefin-based separator 700/39K was used since it is very stable in the very alkaline conditions created by the electrolyte. At this point, it is worth pointing out that a detailed assessment of the separator stability is a crucial step in the assembling of the device.

2.2. Definition of Cell Geometry

Considering that the ultimate objective is to create a 12 V battery that can power a 12 W LED streetlamp for twelve hours, it is crucial to define the capacity of the device. To that end, an approximately 15 A·h capacity battery would be required. However, the geometrical constraints of the lamp limit the battery dimensions to 27 cm × 15 cm × 3.2 cm, restricting the device geometry and number of stacks that can be assembled in series (to meet voltage targets) and in parallel to ensure the required capacity.

Since the objective is to deliver a prototype tailored to meet the application constraints, and because there are no available commercial cases, ideally, a pouch cell, as the one displayed in Figure 1, adapted to the said dimensions could be used. However, the development of the casing requires additional design, developments and optimization to avoid electrolyte leakage, in addition to special equipment and accessories to assemble the casing parts. In this instance, a specially designed battery casing would be required for the specific application, but it is not available at a commercial scale. Moreover, this casing would entail preparing electrodes with an approximate dimension of 28 cm × 14 cm. The effect of such geometry raises another main challenge to bench-scale production: the amount of active material required to cover the conductive substrate would require dedicated laboratory facilities. Furthermore, ensuring that the basic material properties are not altered by scaling up is critical, as this process would greatly impact the final device metrics and, therefore, the number of stacks that would be required to meet capacity and voltage specifications.

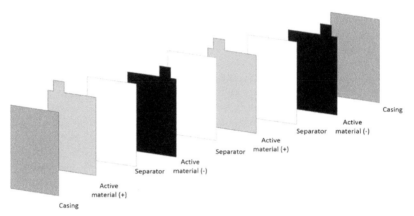

Figure 1. Schematic representation of a pouch cell.

Also, dedicated equipment must be used to measure the electrochemical response of the device. It is important to note that the measuring equipment must reach the intended voltage and current ranges. For example, regular potentiostats are limited to the ±5 A and ±6 V range, thus suitable equipment must be available. Therefore, developing a tailored casing, purchasing the adequate measuring equipment and producing enough active material to assemble the electrodes may consume significant time and must be considered during the planning of the first iteration to produce a prototype. In this specific example and because of the above-mentioned constraints, it was decided to assemble a small device for which the amount of material and existence of commercial casing was ensured. Thus, it was decided to base the prototype on coin cells.

These are commercially available, thus accessible, and can also be easily coupled in series (to increase the operating voltage) and parallel (to increase the output current). To that end, stainless-steel R2032 cells were selected, since they comprise all the desired requirements: commercially accessible, the amount of material introduced leads to a current range that falls within the measurable range of current and potential and the required active material is compatible with the scalability of the synthesis procedure at a bench scale. It is a good starting point to ground the assembling of small prototypes and to identify the constraints raised by the scale-up.

3. Electrochemical Performance of Individual Electrodes

3.1. Electrochemical Performance of Positive Electrode

The active material for the positive electrode presents activity in the positive potential range vs. a saturated calomel electrode (SCE), displaying voltammograms with two quasi-reversible redox peaks, located at 0.35 V and 0.00 V vs. SCE at 5 mV·s^{-1}, as shown in Figure 2. The peaks are displaced with increasing scan rate, as expected for a faradaic dominated electrochemical response.

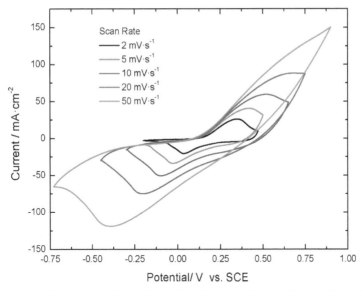

Figure 2. Cyclic voltammetry results for the positive electrode, nickel-cobalt hydroxide, at different applied scan rates.

The capacity of this active material, when tested in a 3 electrodes electrochemical cell is 2.21 C cm^{-2} at 10 mA cm^{-2} with 92% Coulombic efficiency (100·discharge time/charge time), as shown in Figure 3.

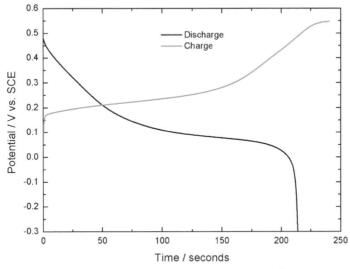

Figure 3. Galvanostatic charge and discharge curves obtained at 10 mA cm^{-2} for the positive electrode, nickel-cobalt hydroxide.

The capacity values obtained at different applied currents have been 2.08 C cm^{-2} at 20 mA cm^{-2}, 2.02 C cm^{-2} at 30 mA cm^{-2}, 1.98 C cm^{-2} at 40 mA cm^{-2}, 1.93 C cm^{-2} at 50 mA cm^{-2}, 1.87 C cm^{-2} at 60 mA cm^{-2} and 1.80 C cm^{-2} at 70 mA cm^{-2}. These results, calculated during the third discharge curve in the −0.3 V to 0.5 V voltage range, exemplify the capacity retention properties of the material, retaining up to 82% of the initial capacity when the applied current density is increased seven times. This is exemplified in Figures 4 and 5.

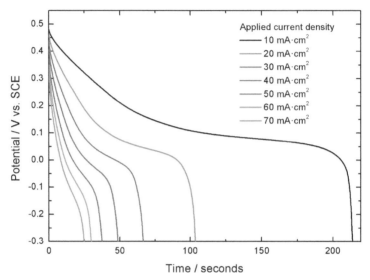

Figure 4. Galvanostatic discharge curves of nickel-cobalt hydroxide at different applied current densities.

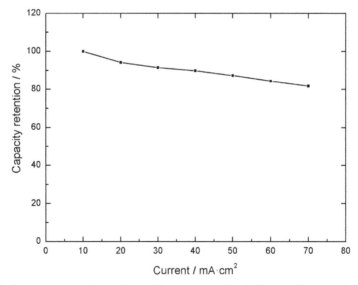

Figure 5. Capacity retention obtained for nickel-cobalt hydroxide calculated at different applied current densities calculated from the third discharge curve in the −0.3 V to 0.5 V vs. SCE voltage interval.

Finally, when the cycling stability of the material is evaluated, 78% of the initial capacity is retained after 2000 cycles at 100 mA cm^{-2} in the −0.3 V to 0.5 V vs. SCE interval, as shown in Figure 6, which is an important achievement for this type of material.

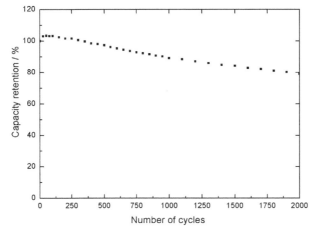

Figure 6. Capacity retention with continuous cycling of the positive electrode material at 100 mA cm^{-2} in the -0.3 V to 0.5 V vs. SCE voltage interval.

In conclusion, these results pinpoint the excellent performance of nickel-cobalt hydroxide on top of carbon paper as potential positive electrode for a hybrid energy storage device. Therefore, in this step, the positive active material, tested in a 3 electrodes electrochemical assembly, is fully validated.

3.2. Electrochemical Performance of Negative Electrode

The electrochemical response of commercially available activated carbon YEC-8A was tested at different scan rates by cyclic voltammetry—Figure 7. As expected for an electrical double-layer capacitive material, the material shows a stable electrochemical response and progressive growth of the measured current with increased scan rates, as required for a carbon-based electrode with optimized performance and cycling stability.

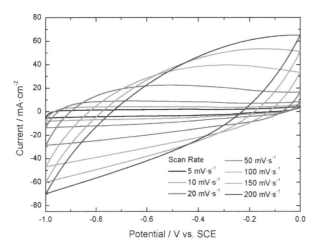

Figure 7. Cyclic voltammetry results obtained for YEC-8A in KOH 6M at different scan rates in the 0.0 V to -1.0 V vs. SCE interval.

The material displays linear discharge curves, in accordance with electrochemical double layer charge storage, yielding 0.61 C cm^{-2} at 10 mA cm^{-2} and a 98% Coulombic efficiency, as displayed in Figure 8.

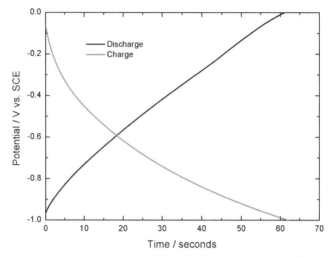

Figure 8. Galvanostatic charge and discharge curves obtained at 10 mA cm^{-2} for the negative electrode, YEC-8A.

Another important characteristic is the electrode electrochemical response under different current loads. In this case, a 71% capacity retention was displayed when the current was varied from 10 to 80 mA cm^{-2} of active material, as exemplified in Figures 9 and 10.

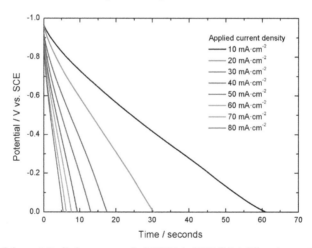

Figure 9. Galvanostatic discharge curves for YEC-8A in KOH 6M at different current densities.

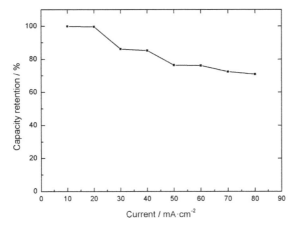

Figure 10. Capacity retention obtained for YEC-8A calculated at different applied current densities calculated from the third discharge curve in the −1.0 V to 0.0 V vs. SCE voltage interval.

Furthermore, an increase in capacity was observed after 2000 cycles, reaching a 104% of the initial capacity, as observed in Figure 11. Therefore, the material presents excellent stability under continuous cycling in the 0.0 V to −1.0 V vs. SCE voltage range. At this stage, the negative electrode material was validated in a 3 electrodes electrochemical cell as well.

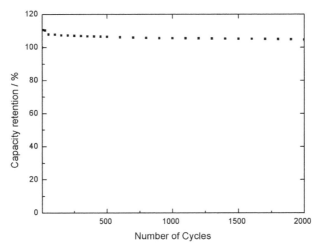

Figure 11. Capacity retention with continuous cycling of the negative electrode material at 10 mA cm^{-2} obtained in the 0.0 V to −1.0 V vs. SCE voltage range.

In conclusion, the positive and the negative materials selected present very promising electrochemical performance and very high stability. When tested individually, they satisfy the crucial requirements for the intended application.

4. Prototype Assembly

The next step is to build the single electrochemical cell: a coin cell. Thus, a 3.2 mm thick SS316 stainless-steel coin cell with 20 mm of diameter was used as casing and current collector. Inside, the coin cell contains two electrodes, the separator and a spring with a metal plate, as exemplified in Figure 12.

Figure 12. Casing parts of coin cells, including one spring, the electrodes and the separator.

During the first step, two 15 cm × 7 cm carbon paper substrates were coated with the active material ink to produce the electrodes. These were then dried and cut into 16 mm diameter disks to match the required dimensions of the coin cell casing. The electrodes were tested with different thicknesses, to account for charge balance in both electrodes, which is critical to ensure the cell optimized response. Best results were obtained for 0.84 mm thickness and 0.48 mm for the negative and positive electrodes thickness, respectively. Between the two electrodes, a 18 mm diameter polyolefin separator, with 0.12 mm thickness, was placed to avoid short-circuiting the cell. At this point, critical issues are related to the assembling process and the sequence of different steps that must be individually optimized. The assembly sequence was the following:

1. Place positive electrode in the positive side contact.
2. Electrolyte dosage.
3. Placement of the separator to ensure coverage of the entire surface of the positive electrode.
4. Electrolyte dosage.
5. Placement of the negative electrode aligned with the positive electrode.
6. Conductive stainless-steel plate to create contact with the negative electrode.
7. Spring.
8. Negative case.
9. Assembling pressure of 800 psi.

There are two key factors identified in the assembly process. First, the pressure used in the production of the coin cell needs to be optimized in advance. In this specific case, an applied pressure of 600 psi did not ensure electrolyte retention and hermetic sealing while 1000 psi would lead to case deformation and risk of short circuiting. Second, the electrolyte dosage had to be optimized. The two-time controlled addition of small volumes of electrolyte (0.05 mL) was required to ensure ionic conductivity within the cell while avoiding leakage during the assembling process. Again, these are steps that require several preliminary trials to ensure that the assembling succeeds.

Finally, for the testing of the coin cells, conductive connectors of stainless steel 304 were attached to the case, as observed in Figure 13, for ease of measuring and to evaluate the electrochemical response of the cell.

Figure 13. Image of an individual coin cell with welded connections prepared to be tested.

The number of springs inside the cell could be adjusted to either one, two or none. Experimental trials showed no difference in terms of electrochemical response under the different possibilities. In conclusion, optimizing the thickness of the electrode, the number of springs, the electrolyte dosage, and the electrode and separator correct dimensioning entailed an iterative optimization process that ensures overcoming any imposed assembling challenge, while respecting safety, reproducibility and performance specifications.

5. Prototype Performance

Initially, the voltage of the cell was limited from 0.0 V to 1.2 V, as shown in Figure 14. Galvanostatic charge-discharge tests were done to evaluate the performance of the cell at different applied currents.

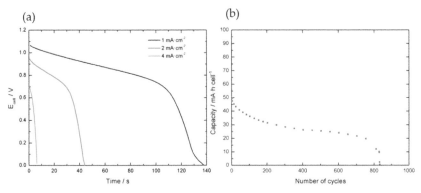

Figure 14. (**a**) Galvanostatic discharge curves for different applied current densities in the 0.0 V to 1.2 V range for a coin cell. (**b**) Cycling stability test for a coin cell in the 0.0 V to 1.2 V voltage range at 1 mA cm^{-2}.

The initial calculated value was around 0.15 mA·h·cell^{-1}, but the cell evidenced faster capacity decay compared to the individual materials and loss of performance after approximately 825 cycles. After the initial tests, it was concluded that the voltage range should be limited to 0.6 V in the lower limit to ensure better cycling stability.

Performance was tested in the more adequate 0.6 V–1.4 V voltage range after polarization at 1.45 V and open circuit potential (which accounts for cell self-discharge) according to industrial standards. Figure 15 displays galvanostatic discharge curves at different applied specific currents in the modified voltage range.

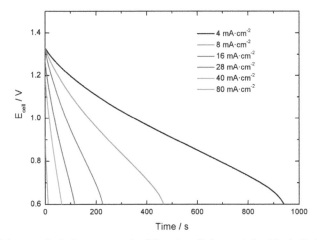

Figure 15. Galvanostatic discharge curves for different applied current densities in the 0.6 V to 1.4 V range for the coin cell.

At this end, it was found that the single cells present around 1 mA·h in a 0.7 V voltage range. An important decay in the capacity response is observed in comparison to the performance of the individual electrode materials, as expected. Indeed, this agrees with the general observations that scaling-up factors may induce important capacity losses. Results of single cells performance are summarized in Table 1.

Table 1. Performance evaluation of the produced coin cells.

Current/mA·cm^{-2}	Capacitance/F·Cell^{-1}	Capacity/mA·h·Cell^{-1}	Capacitance Retention/%
4	5.14	1.04	100
8	4.96	1.03	97
16	4.94	1.00	96
28	4.81	0.89	94
40	4.35	0.71	85
80	2.76	0.24	54

Figure 16 shows a comparison of the charge and discharge curves, which exhibit good symmetry that results in 90% Coulombic efficiency. An evaluation of the ESR (by means of voltage drop between charge and discharge) results in a value of 13 Ω.

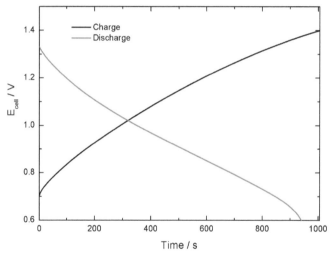

Figure 16. Galvanostatic charge and discharge curves at 1 mA cm^{-2} in the 0.6 V to 1.4 V range for the coin cell.

The tests demonstrate that self-discharge does not seem to be a critical effect, evidencing good charge retention capabilities.

Overall the results show acceptable performance of the prototype, with potential for the envisioned application. However, the high ESR value obtained indicates that the system contains many sources of resistance that negatively impact the overall capacity of the cell. This may be caused by the stainless-steel casing. An important source of resistance may be caused by some corrosion of steel in a strong alkali environment.

It is worth noting that, in practical applications, to adequately study scalability, it would be recommended to use a pouch cell device, in which a corrosion-resistant polymer would have been selected as casing material. Nonetheless, the process described above, in which a coin cell was used, has served to optimize electrode thickness, materials integration, electrode mass balance, active voltage window, current density range, coulombic efficiency and self-discharge of the cell and it is perfectly feasible at laboratory scale. Moreover, it is reproducible and cheap. Results verify the potential application of these materials in hybrid device applications and, during another iteration in the optimization process, more suitable casing would be used.

6. Stack Assembly and Performance

To validate the scale-up potential of the produced coin cells, 5-cell stacks, with a 4.5 operational voltage range, were produced and tested. This process, used to evaluate the impact of scaling up in the performance of the cells, is presented in Figure 17. The number of stacked cells was selected to ensure an active voltage range below the limits imposed by the measuring equipment (6 V). The cells were linked in series to ensure electric conductivity (Figure 17b,c) by means of conductive ink and assembled and insulated by means of non-conductive tape. Finally, their operativity was tested using a 3 V LED.

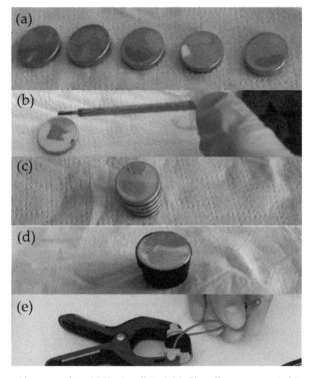

Figure 17. Assembly process for a 4.5 V coin cell stack (**a**). The cells were connected in series (**b,c**) and assembled and insulated by means of non-conductive tape (**d**), and tested with an LED (**e**) that was maintained functional during the expected discharge time of the cell.

Electrochemical characterization of the stack revealed a lower but reasonable Coulombic efficiency, 65%, and a lower current delivered, Figure 18. As expected, the cells assembled in series present limited performance due to increased equivalent series resistance. This is caused by the greater number of interfaces introduced in the device. It is worth mentioning that, in a polymeric casing with the electrodes mounted in series and parallel without the stainless-steel interfaces, this effect would be limited. Nonetheless, at that stage, attention would be required regarding connectors and wiring. Moreover, mechanical integrity of the electrodes would raise other challenges.

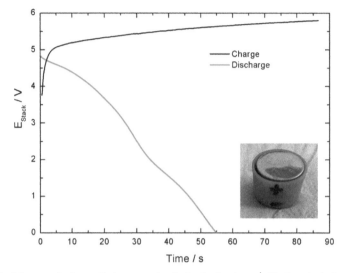

Figure 18. Galvanostatic charge-discharge results obtained at 1 mA cm^{-1}. The inset includes an image of a final stack measured.

Considering the dimensions available for the final device, 29 cm × 15 cm × 3 cm, with 15A·h and 12 V, 12-cell stacks could be used in a similar procedure to the assembly proposed here. To that end, 98 stacks would be required, leading to 1176 coin-cells. Even though stainless-steel cases would not be used, producing such a number of electrodes can be a challenging process for a bench-scale application, in particular, regarding the amount of active material required and the stacks testing and assembly. Thus, the stacks shall be produced in a similar assembling process as a pouch cell, illustrated in Figure 1, ensuring the good contact between the different constituents, mechanical integrity of the individual electrodes, appropriate electrolyte dosage and suitable connections.

Alternative solutions to improve cell performance should be considered. For example, different materials with different electrochemical fingerprints could lead to an increased power response, different synthesis that comply with the scalability required could have been explored and variations of the materials used in this work could have been used, such as different nickel-cobalt ratios or different activated carbons. A significant number of parameters can be adjusted and their influence in the device performance tested. Nonetheless, the case study reported here highlights some important steps in the assembling of a prototype and unveils some of the constraints and challenges that prototyping may face.

7. Materials and Methods

Nickel-cobalt hydroxide was synthesized by co-precipitation. To that end, 5 g of nickel nitrate hexahydrate (Sigma Aldrich) and 10 g of cobalt nitrate hexahydrate (Sigma Aldrich) were dissolved in 100 mL of distilled water. Then, 40 mL of 2M NaOH (Sigma Aldrich) that contained 1.5 mL of H_2O_2 (Merck) were added in the solution forming a green-brown precipitate of Ni-Co(OH)$_2$. The product was aged in solution for 24 h and was then centrifuged at 4000 rpm for 5 min. After centrifugation, it was washed with distilled water and centrifuged again. This process was repeated 6 more times, 3 with ethanol, until neutral pH was achieved. In this way, the removal of impurities was ensured. Approximately 7 g of the material were recovered after the powder was dried at 40 °C for 48 h.

Carbon fiber paper from Toray$^{®}$ was used as conductive substrate, while commercially available activated carbon YEC-8A was purchased from Fuzhou Yihuan Carbon Co. Polyolefin-based separator

700/39K was provided by the Freudenberg group while Stainless-steel 316 R2032 casing cells were purchased from Gelon Lib Co Ltd. and used as current collectors and casing.

Electrochemical measurements were performed by means of an Interface 5000E Potentiostat (Gamry Instruments). Platinum foil of 2.5 cm^2 × 2.5 cm^2 was used as counter electrode while SCE electrode was used as reference in three-electrode configuration for the testing of individual electrodes. KOH 6M was used as alkaline electrolyte in any case.

8. Conclusions

This case study enabled producing a hybrid electrochemical device prototype that serves as basis to design a pre-commercial device. In a future iteration, a pouch cell stacking in polymeric casing shall be developed, to avoid excessive resistive interfaces. Moreover, a specifically designed casing could be developed to accommodate to the geometrical requirements imposed.

Nonetheless, this work evidences the optimization of material integration, when starting from a novel combination of materials to create a prototype and future device. Furthermore, it allowed to optimize electrode thickness, to define the best current collectors, to select a reliable separator and stable aqueous electrolyte, environmentally friendly and operational at high temperatures. Even more, a prototype was built, and its performance investigated, determining Coulombic efficiency, capacity, working potential range, self-discharge, and cycling properties.

The overall resistance of the cell, the biggest limiting factor found, can be circumvented with a polymer-based casing. Thus, the electric resistance built in the cells connected in series can be dramatically reduced. These unitary cells can then be connected in parallel to increase the output current of the cell.

A bench to market process usually lasts around 8–10 years and involves a challenging process of design, product optimization and validation prior commercialization. Nonetheless, the present work can serve as basis to guide researchers on developing simple battery prototypes and evidences the challenges, limitations and potential of scaling-up laboratory created materials.

Author Contributions: A.A.-M., P.A.-C., T.M.e.S., J.C. and M.d.F.M. contributed to the conceptualization of this work, A.A.-M. and P.A.-C. methodology and investigation; A.A.-M., P.A.-C., T.M.e.S. and M.d.F.M. validated the results, A.A.-M. and M.d.F.M., writing—original draft preparation; A.A.-M., P.A.-C., T.M.e.S., J.C. and M.d.F.M., writing—review and editing; M.d.F.M. and J.C., resources.

Funding: This article is a result of the project LLESA, supported by Competitivity and Internationalization Operational Programme (COMPETE 2020), under the PORTUGAL 2020 Partnership Agreement, through the European Regional Development Fund (ERDF). The authors from CQE would also like to acknowledge Fundação para a Ciência e a Tecnologia (FCT) under the funding UID/QUI/00100/2019.

Acknowledgments: We would like to thank Rui Silva and André Mão de Ferro, from C2C-NewCap, for their valuable input and overall aid.

Conflicts of Interest: The authors declare no conflict of interest.

References

1. Bayus, B.L. An Analysis of Product Lifetimes in a Technologically Dynamic Industry. *Manag. Sci.* **2008**, *44*, 763–775. [CrossRef]
2. Héder, M. From NASA to EU: The evolution of the TRL scale in Public Sector Innovation. *Innov. J.* **2017**, *22*, 1–23.
3. United Nations. *Transforming our world: The 2030 Agenda for Sustainable Development*; General Assembly: New York, NY, USA, 25 September 2015; A/RES/70/1; pp. 1–35.
4. Dell, R.M. Batteries-fifty years of materials development. *Solid State Ionics* **2000**, *134*, 139–158. [CrossRef]
5. Janek, J.; Zeier, W.G. A solid future for battery development. *Nat. Energy* **2016**, *1*, 16141. [CrossRef]
6. Aurbach, D.; Mccloskey, B.D.; Nazar, L.F.; Bruce, P.G. Advances in understanding mechanisms underpinning lithium–air batteries. *Nat. Energy* **2016**, *1*, 16128. [CrossRef]
7. Das, S.K.; Lahan, H. Aluminium-ion batteries: Developments and challenges. *J. Mater. Chem. A Mater. Energy Sustain.* **2017**, *5*, 6347–6367. [CrossRef]

Batteries **2019**, *5*, 65

8.   Ochai-Ejeh, F.O.; Madito, M.J.; Momodu, D.Y.; Khaleed, A.A.; Olaniyan, O.; Manyala, N. High performance hybrid supercapacitor device based on cobalt manganese layered double hydroxide and activated carbon derived from cork. *Electrochim. Acta* **2017**, *252*, 41–54. [CrossRef]

9.   Wang, G.; Zhang, L.; Zhang, J. A review of electrode materials for electrochemical supercapacitors. *Chem. Soc. Rev.* **2012**, *41*, 797–828. [CrossRef] [PubMed]

10.   Gören, A.; Costa, C.M.; Silva, M.M.; Lanceros-Méndez, S. State of the art and open questions on cathode preparation based on carbon coated lithium iron phosphate. *Compos. Part B Eng.* **2015**, *83*, 333–345. [CrossRef]

MDPI

St. Alban-Anlage 66

4052 Basel

Switzerland

Tel. +41 61 683 77 34

Fax +41 61 302 89 18

www.mdpi.com

Batteries Editorial Office

E-mail: batteries@mdpi.com

www.mdpi.com/journal/batteries

9 783039 367221